Ergodic Theorems and Related Problems

Ergodic Theorems and Related Problems

V.M. Shurenkov

///VSP///

Utrecht, The Netherlands, 1998

VSP BV
P.O. Box 346
3700 AH Zeist
The Netherlands

First published in 1998

ISBN 90-6764-282-7

Printed in The Netherlands by Ridderprint bv, Ridderkerk.

Valentin Mikhailovich Shurenkov

(1947-1994)

Contents

Preface

It is difficult to find, in the theory of random processes, a term with so many different meanings as 'ergodicity'. In the theory of stationary processes, for example, ergodicity is often identified with metric transitivity. In the theory of Markov processes, the word ergodic is applied to theorems of both the existence of transition probability limits and on the convergence of mean value ratios of these transition probabilities. The word is also used in relation to theorems of convergence of the ratios of the additive functionals' mean values, and for theorems of the convergence of the ratios of the 'time' means in random fields. 'Convergence' can be understood as a convergence in probability, or as mean square convergence.

In addition, there are also 'ergodic theorems' on the convergence of distributions of shifted random processes.

In this monograph, the term 'ergodic' is understood in its original sense, i.e. the one it had when it was first adopted by the theory of random processes from statistical machanics and Boltzmann's theory of gases. By an ergodic theorem, we are referring to any statement about the existence of a mean value with respect to trajectories of a random process taken with respect to time. Here, the author shares the point of view that the problems of the existence of time means, and their equality to the phase means, are interesting without any assumptions about the distribution of the random process.

The book is not intended to be light reading. Even though, to understand the material of the first chapter, it is sufficient to have a general acquaintance with the theory of random processes. To read the second chapter, the reader is assumed to be able to freely use conditional probabilities and to be acquainted with the martingale theory (although this term does not explicitly appear in the book). In the third chapter, the reader is assumed to know the theory of Markov processes and the theory of processes with independent increments given in the second and fourth chapters of the second volume of the book "Theory of random processes" by I. I. Gikhman and A. V. Skorokhod (Moscow, "Nauka", 1973).

The author

Introduction

The following two problems could serve as an introduction to the subject of this book. The first problem is related to the problem of how adequate a mathematical model is for the description of a physical process.

In this problem, let $x_0, x_1, \ldots, x_n \ldots$ be a sequence of states of a certain physical (e.g. mechanical, chemical, biological, etc.) system, $\mathcal{S}$, which evolves in discrete time intervals, t, where $t = 0, 1, 2, \ldots$. To provide a mathematical description of the system $\mathcal{S}$ a transition operator T is used, which acts on the set X of all possible states of the system (called the phase space), such that $x_n = Tx_{n-1}$ and, consequently, $x_n = T^n x_0$. When answering the question of whether this operator T, the choice of which, of course, is substantiated by a sufficiently convincing consideration, is a good mathematical description of the system $\mathcal{S}$, one encounters at least two difficulties. Firstly, it is impossible to determine exactly the states of the system $\mathcal{S}$ (because every measurement inevitably contains an uncertainty) and secondly, it is difficult mathematically to calculate $T^n x$, even if the phase space, X, is relatively simple.

If the uncertainty of the measurement of the phase function $f(x)$ is not systematic, it can be minimized by applying the arithmetical mean $\frac{1}{n}[f(x_0) + f(x_1) + \ldots + f(x_{n-1})]$ for sufficiently large n. However, if n is large, then the expression $\frac{1}{n}[f(x) + f(Tx) + \ldots + f(T^{n-1}x)]$ should not differ much from its limit value;

$$\lim_{n \to \infty} \frac{1}{n} \sum_{k=0}^{n-1} f(T^k x) \tag{1}$$

if, of course, a limit value exists. The situation can be further simplified if this limit does not depend on x. In this case, this limit expression (1) can be written, under fairly general assumptions, as $\int f(x)\mu(dx)$, which may lead to its determination and hence it makes it possible to decide on the adequacy of the system $\mathcal{S}$ and its mathematical description T.

The other problem mentioned above is related to making an estimate of the reliability of a newly designed system if the system is described by random processes ζ_t $(t \geq 0)$ (the more elements the system has, the larger is the random factor in the system's evolution).

One of characteristics determining the reliability of a newly designed system is the fraction of time the processes ζ_t $(t \geq 0)$ spend in the set of states A, i.e.,

$$\lim_{t\to\infty} \frac{1}{t} \int_0^t i_A(\zeta_u)du. \tag{2}$$

Moreover, from the point of view of reliability, the following characteristic of the system, which is more general than expression (2), may be of interest:

$$\lim_{t\to\infty} \frac{1}{t} \int_0^t T_u \xi du \qquad (3),$$

where T_u is an operator that transforms the trajectory of the processes ζ_t $(t \geq 0)$ in u. For example, if $\xi = i_{\tau > c}$, where τ is the time at which the set of the states of failure is attained for the first time, then the limit expression (3) is equal to the proportion of those moments after which the system functioned properly for at least c units of time.

If the limit expression (3) exists with unit probability and is not random, then its value is equal to the mean of $\mathbf{P}^\bullet \xi$, calculated with respect to some probability measure $\mathbf{P}^\bullet$; which, in general, is different from the initial probability. In particular, the limit expression (2) is equal to $\mathbf{P}^\bullet(\zeta_0 \in A)$. Hence, the two problems are reduced to conditions of existence and degeneracy, and to the calculation of the time mean values in (1) for the first problem, and that of the time mean values in (3) for the second. In spite of formal differences, each of them can be written in terms of the other, provided that the phase space X has the structure of a measurable space with a finite measure, the function f given in (1) is measurable, and existence of limit (1) is assumed for most $x \in X$.

The first chapter of this book is devoted to the existence conditions for limits (1) and (3) in precisely such a setting.

In the second chapter the results from the first chapter are employed in the case of random processes with a Markov property.

The final chapter deals entirely with the proof of the ergodic theorem for a very special class of Markov processes, known as a random walk on the semi-axis.

In this book we shall use the following notations, conventions, abbreviations, and definitions. The probability space $(\Omega, \mathcal{F}, \mathbf{P})$ is assumed to contain a sufficient number of subsets in order to define on it all the random processes and variables used in this book.

If ξ and η are random variables and Γ is an event, then, by definition;

$$\mathbf{P}\xi = \int_\Omega \xi d\mathbf{P}, \qquad \mathbf{P}(\xi, \Gamma) = \int_\Gamma \xi d\mathbf{P}, \qquad \mathbf{P}\xi\eta = \mathbf{P}(\xi\eta).$$

We will say that a measure which is generated by a random process, defined for all functions taking values in the phase space of this process and defined on the set of time parameters of the process, is called the distribution of this process.

To each subset, D, of some space, X, there is a function, i_D, on X, called the indicator of the set D, which is equal to unity on D and is zero outside of D. The set which is the complement of D in X will be denoted by $X \rceil D$ or, if it is clear what the set X is, then it will be denoted just by $\rceil D$.

All measures on the real line R_1, encountered in this book, are considered to be defined on the Borel σ-algebra, $\mathcal{B}$. By a measure, we understand that it is a *nonnegative* countably additive function of sets. The symbol $\int_a^b$ always denotes integration over the interval $[a, b]$.

To express the minimum of two numbers a and b, two types of notations are used: $\min(a, b)$ and $a \wedge b$.

The left and right limits of a function f at a point a are denoted by $f(a-)$ and $f(a+)$ and the upper and lower limits are denoted by $\overline{\lim}$ and $\underline{\lim}$, respectively.

For a matrix A with elements a_{ij}, $i, j = 1, \dots, d$, we shall use the notation $\|a_{ij}\|_{i,j=1}^d$. A vector x having the coordinates x_i, where $i = 1, \dots, d$ will be denoted by $(x_i)_{i=1}^d$. The scalar product $\langle x, y\rangle$ of vectors $x = (x_i)_{i=1}^d$ and $y = (y_i)_{i=1}^d$ equals $\sum_{i=1}^d x_i y_i$, and the tensor product, $x \times y$, is the matrix $\|x_i y_j\|_{i,j=1}^d$.

By the Perron root of a matrix with nonnegative off-diagonal elements, the real nonnegative eigenvalue of the matrix has a maximal real part.

A matrix $A = \|a_{ij}\|_{i,j=1}^d$ is called indecomposable if, for any nontrivial subset $\mathcal{J} \subset \{1, \dots, d\}$ there exists indices $i \in \mathcal{J}$ and $j \notin \mathcal{J}$ such that $a_{ij} \neq 0$.

The eigenvalues of the transposition, $A^{\top}$ of the matrix A, are called left eigenvalues of the matrix A. The identity matrix will be denoted by E, and the vector with all the coordinates equal to unity will be denoted by $\mathbf{1}$.

The book is subdivided into different sections. Each subsection contains only one theorem, corollary, lemma, proposition, or example. When a theorem, corollary, lemma, etc. is referred to, the number of the corresponding subsection is given. For example, the second corollary of the first theorem would be referred to as corollary seven.

Chapter 1

Stationary processes

Any ergodic theorem in the theory of random processes, i.e. a theorem about the conditions of existence or degeneracy of time means, inevitably leads to the notion of a stationary process. This process is obtained by replacing the probability measure on the main probability space $\mathbf{P}$ with the measure $\mathbf{P}^\bullet$, which determines the values of the time means. The first fundamental results concerning ergodic theory were obtained for stationary processes (sequences), i.e. for transformations that preserve the measure. This provides a method to prove ergodic theorems by first studying the measure $\mathbf{P}^\bullet$; if, of course, one can find (or, more often, guess) an analytical expression for $\mathbf{P}^\bullet$ in terms of $\mathbf{P}$. This may involve notions that do not belong to the theory of random processes. Thus the fundamental results are given in terms of the time means given in (1) in the preface.

1.1 Convergence and invariance

1. Let a phase space X have the structure of a measurable space, i.e. a σ-algebra $\mathcal{G}$ of subsets exists on which a finite measure μ is defined.

A measurable mapping, T, that maps the measurable space $(X, \mathcal{G})$ into itself is called *ergodic* if, for any bounded numerical measurable function f, the limit

$$\lim_{n\to\infty} \frac{1}{n} \sum_{k=0}^{n-1} f(T^k x) \tag{4}$$

exists almost everywhere with respect to μ and does not depend on x.

Such a definition excludes different pathologies that might occur (for example, uncountably additive 'measures'); maintaining, on the other hand, a sufficient degree of generality.

2. If limit (4) exists almost everywhere with respect to μ, then its value $f^\bullet$ defines a finite measure $\mu^\bullet$ on $\mathcal{G}$ such that

$$\int f^\bullet(x)\mu(dx) = \int f(x)\mu^\bullet(dx). \tag{5}$$

Indeed, by the Lebesgue bounded convergence theorem;

$$\int f^\bullet(x)\mu(x) = \lim_{n\to\infty}\int f(x)\mu_n(dx), \tag{6}$$

where $\mu_n(A) = \frac{1}{n}\sum_{k=0}^{n-1}\mu(T^{-k}A)$ and $A \in \mathcal{G}$, $T^{-k}A = \{x : T^k x \in A\}$. The limit on the right hand side of (6) exists for any bounded measurable function f. Hence, the equation $\mu^\bullet(a) = \lim_{n\to\infty}\mu_n(a)$ allows the definition of an additive nonnegative bounded function $\mu^\bullet$ on the σ-algebra $\mathcal{G}$. Because an arbitrary measurable bounded function can be uniformly approximated by measurable functions, which take a finite number of values, it follows that (5) is true. The countable additivity of $\mu^\bullet$ is equivalent to the relation:

$$\sup_n \mu_n(A_k) \underset{k\to\infty}{\longrightarrow} 0 \tag{7}$$

for an arbitrary monotone sequence of sets from $\mathcal{G}$, $A_1, A_2, \ldots$, which have an empty intersection. This can be proved by contradiction. Indeed, if $\sup_n \mu_n(A_k) \geq \epsilon$ for some ϵ such that $\epsilon > 0$, then $\mu_{n_1}(A_1) \geq \epsilon$ for some natural n_1. The countable additivity of the measure μ_{n_1} allows the choice of a number $n_2 > n_1$, for which $\mu_{n_1}(A_{n_2}) \leq \epsilon/2$. By using n_2, n_4 and n_3 can be chosen such that $n_4 > n_3 > n_2$, $\mu_{n_3}(A_{n_2}) \geq \epsilon$ and $\mu_{n_2}(A_{n_4}) \leq \epsilon/2$. In general, there exists an increasing sequence of numbers $n_1, n_2, \ldots$ such that;

$$\mu_{n_{2k+1}}(A_{n_{2k}}) \geq \epsilon \qquad \text{and} \qquad \mu_{n_{2k+1}}(A_{n_{2k+2}}) \leq \frac{\epsilon}{2},$$

and, consequently,

$$\mu_{n_{2k+1}}(A_{n_{2k}} \rceil A_{n_{2k+2}}) \geq \frac{\epsilon}{2}, \tag{8}$$

where the sets $B_k = A_{n_{2k}} \rceil A_{n_{2k+2}}$ do not intersect.

The set C, where $C = \cup_k B_k$ can be written as a sum of a countable number of nonintersecting sets, $C_j = \cup_r B_{k_{r,j}}$, each of which contains an infinite number of the sets B_k. For example, $C_j = \cup_{r=1}^{\infty} B_{p_j^r}$, $j = 1, 2, \ldots$, $C_0 = C \rceil \cup_{j=1}^{\infty} C_j$, where p_k is the kth natural number. From Equation (8), the following can be obtained:

$$\begin{aligned}
\mu^\bullet(C_j) &= \lim_{n\to\infty}\mu_n(C_j) \\
&= \lim_{r\to\infty}\mu_{n_{2k_{r,j}+1}}(C_j) \geq \overline{\lim_{r\to\infty}}\,\mu_{n_{2k_{r,j}+1}}(B_{k_{r,j}}) \geq \epsilon/2
\end{aligned}$$

which is a contradiction since $\mu^\bullet$ is finite. This demonstrates that Equation (7) is valid and, hence, the measure $\mu^\bullet$ is countably additive.

3. The measure $\mu^\bullet$ and the transformation T are connected by the following invariance equation;

$$\mu^\bullet(T^{-1}A) = \mu^\bullet, \tag{9}$$

which follows from the identity;

$$\mu_n(T^{-1}A) = \frac{n+1}{n}\mu_{n+1}(A) - \frac{1}{n}\mu(A).$$

Equation (9) is equivalent to the following:

$$\int f(Tx)\mu^\bullet(dx) = \int f(x)\mu^\bullet(dx). \tag{10}$$

Moreover, if a measurable set in $\mathcal{G}$, A, is invariant with respect to T, i.e. $T^{-1}A = A$, then $\mu^\bullet(A) = \mu(A)$.

Hence, the existence of a measure that is invariant in T, $\mu^\bullet$, which coincides with μ on the σ-algebra $\mathcal{I}$ of the sets invariant with respect to T, is a necessary condition for the time means;

$$\frac{1}{n}\sum_{k=0}^{n-1} f(T^k x) \tag{11}$$

to converge almost everywhere with respect to measure μ for an arbitrary bounded measurable function f.

Conversely, let a transformation T preserve a finite measure $\mu^\bullet$. Then, by the Birkhoff theorem (Corollary 6 here) the means given by (11) converge almost everywhere with respect to $\mu^\bullet$. The convergence set is invariant. So, if the contraction of the measure μ to the σ-algebra $\mathcal{I}$ is absolutely continuous with respect to the same contraction of the measure $\mu^\bullet$, then the convergence takes place almost everywhere with respect to μ. If, in addition, the contraction of μ to $\mathcal{I}$ is absolutely continuous with respect to the contraction of μ to $\mathcal{I}$, then the degeneracy of limit (4) is valid almost everywhere (with respect to μ) implies that the degeneracy is valid almost everywhere with respect to $\mu^\bullet$, hence limit (4) will be equal to $\int f(x)\mu^\bullet(dx)$.

4. The basis for this and the following analysis is the Birkhoff theorem, which, in its turn, is a particular case of the following statement. The following statement also covers the case when the invariance equation given in (9) has a σ-finite solution but cannot be solved in terms of finite measures.

Theorem. *Let a measurable transformation T of a measurable space $(X, \mathcal{G}, \mu)$ preserve the measure μ. Then, if functions f and g, measurable with respect to $\mathcal{G}$, are integrable and:*

$$g(x) > 0, \qquad \sum_{k=0}^{\infty} g(T^k x) = \infty \tag{12}$$

almost everywhere with respect to μ, then the limit;

$$\lim_{n\to\infty} \frac{\sum_{k=0}^{n-1} f(T^k x)}{\sum_{k=1}^{n-1} g(T^k x)} = f^\bullet(x)$$

exists almost everywhere with respect to μ and:

$$\int f^\bullet(x) g(x)\mu(dx) = \int f(x)\mu(dx).$$

A proof of this theorem can be obtained by successive use of the following auxiliary statement, known as the maximal ergodic theorem:

5. **Lemma.** *Suppose that the conditions given in the above mentioned theorem hold, then the function f is integrable with respect to the measure μ, and*

$$\sup_{n\geq 1} \sum_{k=0}^{n-1} h(T^k x) \geq 0$$

is true almost everywhere with respect to μ. Then;

$$\int h(x)\mu(dx) \geq 0. \tag{13}$$

The proof follows from the identity;

$$s_{n+1}(x) = h(x) + s_n^+(Tx),$$

where $s_n(x) = \max_{1\leq j\leq n} \sum_{k=0}^{j-1} h(T^k x)$, $s_n^+(x) = \max\{0, s_n(x)\}$, and the inequality $s_n(x) \leq s_{n+1}(x)$. This implies that $h(x) \geq s_n(x)$ and, consequently;

$$\int_{\{s_n\geq 0\}} h(x)\mu(dx) \geq \int_{\{s_n\geq 0\}} s_n(x)\mu(dx) - \int_{\{s_n\geq 0\}} s_n^+(Tx)\mu(dx).$$

The first integral in the right hand side of this relation is equal to $\int s_n^+(x)\mu(dx)$, and the second integral is not greater than $\int s_n^+(Tx)\mu(dx)$, which in its turn is equal to $\int s_n^+(x)\mu(dx)$ because the measure μ is invariant (the integrability of s_n is implied by the invariance of the measure μ and the integrability of h). Hence;

$$\int_{\{s_n\geq 0\}} h(x)\mu(dx) \geq 0.$$

The limit as $n\to\infty$ is taken, and (13) is thus obtained.

Further, let $f^*(x)$ and $f_*(x)$ be the upper and lower limits respectively of the expression $\sum_{k=0}^{n-1} f(T^k x)/(\sum_{k=0}^{n-1} g(T^k x))$ as $n\to\infty$. Then, according to (12), $f^*(Tx) = f^*(x)$. Indeed;

$$f^*(x) = \lim_{m\to\infty} \sup_{n\geq m} \frac{\sum_{k=0}^{n-1} f(T^k x)}{\sum_{k=0}^{n-1} g(T^k x)} = \lim_{m\to\infty} \sup_{n\geq m} \frac{\sum_{k=0}^{n} f(T^k x)}{\sum_{k=0}^{n} g(T^k x)}$$

and, consequently;

$$\begin{aligned} f^*(Tx) &= \lim_{m\to\infty} \sup_{n\geq m} \frac{\sum_{k=1}^n f(T^k x)}{\sum_{k=1}^n g(T^k x)} \\ &= \lim_{m\to\infty} \sup_{n\geq m} \left\{ \frac{\sum_{k=0}^n g(T^k x)}{\sum_{k=1}^n g(T^k x)} \cdot \frac{\sum_{k=0}^n f(T^k x)}{\sum_{k=1}^n g(T^k x)} - \frac{f(x)}{\sum_{k=1}^n g(T^k x)} \right\}. \end{aligned}$$

Because $\sum_{k=0}^{\infty} g(T^k x) = \infty$ with respect to μ for almost all $x \in X$, then it is true almost everywhere with respect to μ:

$$\frac{\sum_{k=0}^n g(T^k x)}{\sum_{k=1}^n g(T^k x)} \underset{n\to\infty}{\longrightarrow} 1$$

and:

$$\frac{f(x)}{\sum_{k=1}^n g(T^k x)} \to 0.$$

The upper limit of the expression in the brackets, as n is taken to ∞, is equal to:

$$\overline{\lim_{n\to\infty}} \frac{\sum_{k=0}^n f(T^k x)}{\sum_{k=0}^n g(T^k x)} = f^*(x)$$

almost everywhere with respect to μ. Hence, if $\chi(x)$ is the indicator function of the set $\{a_1 < f^* < a_2\}$, then $\chi(x)[f(T^k x) - a_1 g(T^k x)] = h_1(T^k x)$ and $\chi(x)[a_2 g(T^k x) - f(T^k x)] = h_2(T^k x)$ are true, where $h_1 = \chi(f - a_1 g)$ and $h_2 = \chi(a_2 g - f)$.

Also, if $\chi(x) = 1$, then $\sup_{n\geq 1} \left[\sum_{k=0}^{n-1} f(T^k x) / \sum_{k=0}^{n-1} g(T^k x)\right] > a_1$ and $\inf_{n\geq 1} \left[\sum_{k=0}^{n-1} f(T^k x) / \sum_{k=0}^{n-1} g(T^k x)\right] < a_2$. Consequently, $\sup_{n\geq 1} \sum_{k=0}^{n-1} h_1(T^k x) \geq 0$ and $\sup_{n\geq 1} \sum_{k=0}^{n-1} h_2(T^k x) \geq 0$ almost everywhere with respect to μ. The integrability of h_1 and h_2 is implied by the integrability of f and g. By Lemma 5;

$$\int h_1(x)\mu(dx) \geq 0 \qquad \text{and} \qquad \int h_2(x)\mu(dx) \geq 0, \tag{14}$$

or;

$$\begin{aligned} a_1 \int_{\{a_1 < f^* < a_2\}} g(x)\mu(dx) &\leq \int_{\{a_1 < f^* < a_2\}} f(x)\mu(dx) \\ &\leq a_2 \int_{\{a_1 < f^* < a_2\}} g(x)\mu(dx). \end{aligned}$$

These inequalities are equivalent to the following:

$$\begin{aligned} a_1 \int_{\{a_1 \leq f^* < a_2\}} g(x)\mu(dx) &\leq \int_{\{a_1 \leq f^* < a_2\}} f(x)\mu(dx) \\ &\leq a_2 \int_{\{a_1 \leq f^* < a_2\}} g(x)\mu(dx). \end{aligned} \tag{15}$$

If $\chi(x)$ is the indicator function of the set $\{f^* \geq a_1\}$, then the inequalities containing $h_1(x)$ will still be valid. This, in particular, is true for the first inequality of (14), which implies that:

$$a_1 \int_{\{a_1<f^*\}} g(x)\mu(dx) \leq \int_{\{a_1<f^*\}} f(x)\mu(dx).$$

By taking the limit $a_1 \to \infty$, then $f^* < \infty$ is true almost everywhere. Similarly, if $\chi(x)$ is the indicator function of the set $\{f^* < a_2\}$, then all inequalities containing $h_2(x)$ remain valid and the second inequality of (14) yields:

$$\int_{\{f^*<a_2\}} f(x)\mu(dx) \leq a_2 \int_{\{f^*<a_2\}} g(x)\mu(dx),$$

which, if a_2 is taken to ∞, this implies that $f^* > -\infty$ is true almost everywhere with respect to μ.

Thus, if all of the inequalities given in (15), taking $a_1 = (k-1)/2^n$ and $a_2 = k/2^n$ for all integers k, then:

$$\sum_{k=-\infty}^{\infty} \frac{k-1}{2^n} \int_{\{(k-1)/2^n \leq f^* < k/2^n\}} g(x)\mu(dx) \leq \int f(x)\mu(dx)$$

$$\leq \sum_{k=-\infty}^{\infty} \frac{k}{2^n} \int_{\{(k-1)/2^n \leq f^* < k/2^n\}} g(x)\mu(dx).$$

Taking the limit of n to ∞, the equation:

$$\int f^*(x)g(x)\mu(dx) = \int f(x)\mu(dx)$$

is obtained, which, if f is replaced by $-f$, becomes:

$$\int f_*(x)g(x)\mu(dx) = \int f(x)\mu(dx).$$

These equations can only hold simultaneously if $f_* = f^*$ almost everywhere with respect to μ.

6. A particular case of this theorem, when $\mu(X) < \infty$ and $g(x) \equiv 1$, is important enough to be seperately formulated.

Corollary. If a measurable transformation T of a measurable space $(X, \mathcal{G}, \mu)$ preserves the finite measure μ and a measurable (with respect to $\mathcal{G}$) function f, is integrable, then the limit;

$$\lim_{n\to\infty} \frac{1}{n} \sum_{k=0}^{n-1} f(T^k x) = f^*(x)$$

exists almost everywhere with respect to μ and:

$$\int f^*(x)\mu(dx) = \int f(x)\mu(dx).$$

Using the language of the theory of random processes, this theorem can be restated as follows.

7. A sequence of random variables, $\xi_0, \xi_1, \dots, \xi_n, \dots$, is called *stationary* if its distribution is equal to the distribution of the translated sequence, $\xi_1, \xi_2, \dots, \xi_{n+1}, \dots$.

Corollary. If a sequence ξ_n, $n = 0, 1, 2, \dots$, is stationary and $\mathbf{P}|\xi_0| < \infty$, then the limit;

$$\lim_{n\to\infty} \frac{1}{n} \sum_{k=0}^{n-1} \xi_k = \xi^\bullet \tag{16}$$

almost certainly exists and $\mathbf{P}\xi_0^\bullet = \mathbf{P}\xi_0$.

Proof. Let X be the space of the sequences of numbers $x = (x_0, x_1, \dots, x_n, \dots)$ with the σ-algebra $\mathcal{G}$, which is generated by the sets $\{x \in X : x_k < c\}$ for $k = 0, 1, 2, \dots$, $c \in \mathbf{R}^1$. The equation $\mu\{x \in X : x_0 < c_0, x_1 < c_1, \dots, x_n < c_n\} = \mathbf{P}\{\xi_0 < c_0, \xi_1 < c_1, \dots, \xi_n < c_n\}$ defines, on $\mathcal{G}$, a finite measure μ. μ is invariant because it is stationary with respect to the translation, T, of the sequence $x = (x_0, x_1, \dots)$: $Tx = (x_1, x_2, \dots)$. The function $f(x) = x_0$ is measurable with respect to $\mathcal{G}$, integrable with respect to μ and $f(T^k x) = x_k$. For such a choice of X, $\mathcal{G}$, μ, and f, the probability that limit (16) exists is equal to the value that the measure μ takes, in the set of those $x \in X$, for which limit (4) exists (by Corollary 6, this probability is equal to unity). The mean value of limit (16) is equal to the integral of limit (4) with respect to the measure μ, which, by Corollary 6, is equal to $\int f(x)\mu(dx) = \int v x_0 \mu(dx) = \mathbf{P}\xi_0$.

1.2 Translation operators

8. The ergodic theory of random variables deals with a certain kind of transformation: the transformations, T_t, of translation of random variables and elementary events. These transformations can be defined for an arbitrary random process, $\mathfrak{z}_t$ (where $t \geq 0$), in an arbitrary phase space $(E, \mathcal{A})$ and for the set of trajectories stable with respect to the translations. The latter implies that for an arbitrary $\omega \in \Omega$ and any $t > 0$ there exists an $\omega' \in \Omega$ such that $\mathfrak{z}_{s+t}(\omega) = \mathfrak{z}_s(\omega')$ for all $s \geq 0$. Every such mapping T is automatically measurable with respect to the σ-algebra $\mathcal{M}$ generated by the process $\mathfrak{z}_t$ $(t \geq 0)$. Indeed, the σ-algebra $\mathcal{M}$ is also generated by the events $\Gamma = \{\omega : \mathfrak{z}_s(\omega) \in B\}$ for all $s \geq 0$ and $B \in \mathcal{A}$. With respect to the mapping T_t, the inverse image $T_t^{-1}\Gamma$ of such an event Γ is equal to $\{\omega : \mathfrak{z}_s : (T_t\omega) \in B\} = \{\omega : \mathfrak{z}_{s+t}(\omega) \in B\}$, and consequently, if $\Gamma \in \mathcal{M}$, then the set $T_t^{-1}\Gamma$ belongs to $\mathcal{M}$ and does not depend on a particular choice of the mapping T_t.

Similarly, if a random variable ξ is measurable with respect to $\mathcal{M}$, then the equation $T_t\xi(\omega) = \xi(T_t\omega)$ defines uniquely an measurable variable, $T_t\xi$, with respect to $\mathcal{M}$. If ξ is an indicator function of an event Γ from $\mathcal{M}$, then $T_t\xi$ is also an indicator function of the measurable event, $T_t^{-1}\Gamma$, with respect to $\mathcal{M}$.

The operators T_t of the translation of random variables, defined in just such a way as above, commute with taking Borel functions of several variables and with the operation of taking limits, i.e. if φ is a Borel function of n real variables and is a sequence of measurable variables, with respect to $\mathcal{M}$, and $\xi_1, \xi_2, \ldots$ converges everywhere on Ω, thens $T_t\varphi(\xi_1, \ldots, \xi_n) = \varphi(T_t\xi_1, \ldots, T_t\xi_n)$, and $T_t \lim_{n\to\infty} \xi_n = \lim_{n\to\infty} T_t\xi_n$. Moreover, $T_t\mathfrak{z}_s = \mathfrak{z}_{s+t}$ and $T_tT_s = T_{s+t}$.

The translation operators determine the σ-algebra $\mathcal{T}$ of invariant events from $\mathcal{M}$; i.e. $\Gamma \in \mathcal{T}$ if and only if $\Gamma \in \mathcal{M}$ and $T_t^{-1}\Gamma = \Gamma$ for all $t \geq 0$. Similarly a random variable that is measurable with respect to $\mathcal{M}$, ξ, will be measurable with respect to $\mathcal{T}$ if and only if $T_t\xi = \xi$ for all $t \geq 0$.

9. Any well posed problem about the existence and degeneracy of the time means given in (3), necessarily involves the measurability of the mapping $(t, \omega) \to T_t\xi(\omega)$ with respect to the product $\mathcal{B}_+ \times \mathcal{M}$ of the σ-algebra $\mathcal{B}_+$ of Borel subsets of the real half-axes $\mathbf{R}^1_+ = [0, \infty)$ and the σ-algebra $\mathcal{M}$. The measurability of this mapping is equivalent to the measurability of the mapping $(t, \omega) \to \mathfrak{z}_t(\omega)$ of the measurable space $(\mathbf{R}^1_+ \times \Omega, \mathcal{B} \times \mathcal{M})$ into the measurable space $(E, \mathcal{A})$. If such a measurability exists, then the random process $\mathfrak{z}_t$ $(t \geq 0)$ is called *measurable*.

A measurable random process $\mathfrak{z}_t$ $(t \geq 0)$ is called *ergodic* if the limit:

$$\lim_{t\to\infty} \frac{1}{t} \int_0^t T_u\xi du \tag{17}$$

exists and is almost certainly degenerate for any bounded measurable (with respect to $\mathcal{M}$) variable ξ. Limit (17), if it exists, can almost certainly be chosen so it is a measurable variable with respect to $\mathcal{T}$. Moreover, for a measurable ξ (with respect to $\mathcal{T}$), the limit is equal to ξ. Hence, the ergodicity of a process implies that the limits of the time means given in (17) almost certainly exist and that the σ-algebra $\mathcal{T}$ of invariant events is trivial.

10. A process $\mathfrak{z}_t$ $(t \geq 0)$ is called *stationary* if its distribution coincides, for any $s > 0$, with the distribution of the translated process $\mathfrak{z}_{s+t}$ $(t \geq 0)$. This is equivalent to the invariance of the initial probability measure $\mathbf{P}$ on $\mathcal{M}$ with respect to the translation operators T_s; i.e. $\mathbf{P}(T_s^{-1}\Gamma) = \mathbf{P}(\Gamma)$, where $\Gamma \in \mathcal{M}$. If the limits given in (17) exist and are almost certainly degenerate, then the values $\mathbf{P}^\bullet\xi$ define, on $\mathcal{M}$, a probability $\mathbf{P}^\bullet$, which is invariant with respect to the translation operators T_s $(s \geq 0)$; i.e. $\mathbf{P}^\bullet T_s\xi = \mathbf{P}^\bullet\xi$, which coincides with the initial probability $\mathbf{P}$ on events from $\mathcal{T}$. Because the existence of limit (17) is a measurable event with respect to $\mathcal{T}$, and the replacement of the initial probability space $(\Omega, \mathcal{F}, \mathbf{P})$ with $(\Omega, \mathcal{M}, \mathbf{P}^\bullet)$ makes the process $\mathfrak{z}_t$ $(t \geq 0)$ stationary, stationary random variables are a natural object for a preliminary study of the convergence of the time mean values.

11. Theorem. *If a process, $\mathfrak{z}_t$ $(t \geq 0)$, is stationary and a measurable (with respect to $\mathcal{M}$) variable, ξ, is integrable, then:*

$$\lim_{t\to\infty} \frac{1}{t} \int_0^t T_u \xi du = \mathbf{P}(\xi|\mathcal{T}) \tag{18}$$

is almost certainly valid.

Proof. The limit on the left hand side exists if and only if the limit of the sequence;

$$\frac{1}{n} \int_0^n T_u \xi du, \qquad n = 1, 2, \ldots \tag{19}$$

exists. Indeed, let $\xi \geq 0$ (the general case can be reduced to this case) and $n \leq t < n+1$. Then:

$$\frac{1}{n+1} \int_0^n T_u \xi du \leq \frac{1}{t} \int_0^t T_u \xi du \leq \frac{1}{n} \int_0^{n+1} T_u \xi du$$

and the convergence of the sequence given in (19) implies the existence of the limit in the left hand side of equation (18). This, together with the measurability of the process $\mathfrak{z}_t$ $(t \geq 0)$ and the Fubini theorem, implies that the event is measurable with respect to $\mathcal{M}$ (given that the limit in the left hand side of (18) exists). The sequence $\xi_n = \int_n^{n+1} T_u \xi du$, $n = 0, 1, \ldots$, is thus stationary and $\mathbf{P}|\xi_0| \leq \int_0^1 \mathbf{P} T_u |\xi| du = \mathbf{P}|\xi|$.

According to Corollary 7, the limit;

$$\lim_{n\to\infty} \frac{1}{n} \sum_{k=0}^{n-1} \xi_k = \lim_{n\to\infty} \frac{1}{n} \int_0^n T_u \xi du = \xi^\bullet$$

almost certainly exists and $\mathbf{P}\xi^\bullet = \mathbf{P}\xi_0 = \mathbf{P}\xi$. The random variable $\xi^\bullet$ is defined and invariant on a measurable set with respect to $\mathcal{T}$, at which the limit given in (18) exists. Consequently, if the variable $\xi^\bullet$ is defined to be zero on the complement of this set, then it becomes measurable with respect to $\mathcal{T}$. Let a measurable variable η, with respect to $\mathcal{T}$, be bounded. Then, as it has been shown that;

$$\lim_{n\to\infty} \frac{1}{t} \int_0^t T_u(\eta\xi) du = \eta \lim_{t\to\infty} \frac{1}{t} \int_0^t T_u \xi du = \eta\xi^\bullet \quad \text{and} \quad \mathbf{P}\eta\xi^\bullet = \mathbf{P}\eta\xi.$$

This implies that $\xi^\bullet = \mathbf{P}(\xi|\mathcal{T})$.

12. The situation when the measure $\mathbf{P}$ on $\mathcal{M}$ is preserved by random translation transformations T_τ;

$$T_\tau \xi(\omega) = \begin{cases} \xi(T_{\tau(\omega)}\omega), & \text{if } \tau(\omega) < \infty \\ 0, & \text{if } \tau(\omega) = \infty, \end{cases}$$

where τ is a nonnegative and measurable, with respect to $\mathcal{M}$, random variable. Because the process $\mathfrak{z}_t$ $(t \geq 0)$ is measurable, then the variable $T_\tau \xi$ is measurable with respect to $\mathcal{M}$ for any measurable (with respect to $\mathcal{M}$) variable ξ and any nonnegative measurable (with respect to $\mathcal{M}$) τ so that the variables $T_\tau^2\xi$, $T_\tau^3\xi$, ... , are defined and measurable with respect to $\mathcal{M}$. The kth iteration of the operator T_τ can be replaced by the operator T_{τ_k}, which is the translation of τ_k, i.e. $T_\tau^k = T_{\tau_k}$, where $\tau_0 = 0$, $\tau_k = \tau + T_\tau \tau_{k-1}$, $k = 1, 2, \ldots$.

Theorem. *Let a transformation T_τ preserve the probability $\mathbf{P}$ on $\mathcal{M}$ and:*

$$\mathbf{P}(\tau > 0) = 1, \qquad \mathbf{P}\tau < \infty. \tag{20}$$

If $\mathbf{P}\int_0^\tau T_u|\xi|du < \infty$, then:

$$\lim_{t\to\infty} \frac{1}{t} \int_0^t T_u\xi du = \mathbf{P}^\bullet(\xi|\mathcal{T}) \tag{21}$$

almost certainly is valid, where;

$$\mathbf{P}^\bullet\xi = \frac{1}{\mathbf{P}\tau}\mathbf{P}\int_0^\tau T_u\xi du. \tag{22}$$

Proof. Equation (22) determines, on $\mathcal{M}$, a probability measure $\mathbf{P}^\bullet$, which is invariant with respect to the translation operators T_t $(t \geq 0)$. Indeed;

$$\begin{aligned}
\int_0^\tau T_u T_t\xi du &= \int_0^\tau T_{u+t}\xi du = \int_t^{\tau+t} T_u\xi du \\
&= \int_0^\tau T_u\xi du + \int_\tau^{\tau+t} T_u\xi du - \int_0^t T_u\xi du \\
&= \int_0^\tau T_u\xi du + T_\tau \int_0^t T_u\xi du - \int_0^t T_u\xi du
\end{aligned}$$

and the probability $\mathbf{P}^\bullet$ is invariant with respect to T_t, because the probability $\mathbf{P}$ is invariant with respect to T_τ. Moreover, if $\Gamma \in \mathcal{T}$, then $\mathbf{P}^\bullet(\Gamma) = \frac{1}{\mathbf{P}\tau}\mathbf{P}(\tau, \Gamma)$ and, as equation (20) shows, contractions of the probabilities $\mathbf{P}$ and $\mathbf{P}^\bullet$ to the σ-algebra $\mathcal{T}$ are absolutely continuous with respect to each other. Because the set for which both the limit exists and the left hand side of equation (21) is equal to the expression in the right hand side of equation (21), is a measurable event with respect to $\mathcal{T}$, then the statement of Theorem 12 thus follows from Theorem 11.

13. Theorem. *Suppose that the conditions of Theorem 12 are valid, except for condition (20), and $0 < \tau < \infty$ is almost certainly valid. Then:*

$$\lim_{t\to\infty} \frac{\int_0^t T_u\xi du}{\int_0^t T_u\eta du} = \frac{\mathbf{P}\left(\int_0^\tau T_u\xi du|\mathcal{T}\right)}{\mathbf{P}\left(\int_0^\tau T_u\eta du|\mathcal{T}\right)} \tag{23}$$

is almost certainly valid if $\mathbf{P}\int_0^\tau T_u(|\xi| + |\eta|)du < \infty$ and $\mathbf{P}\left(\int_0^\tau T_u\eta du|\mathcal{T}\right) \neq 0$ is almost certainly valid also.

One could prove this theorem in the same way as the previous one by making use of Theorem 4. However, to obtain a better understanding, it might be useful to carry out the following simple argument.

The sequence;

$$\xi_k = \int_{\tau_k}^{\tau_{k+1}} T_u \xi du = T_\tau^k \int_0^\tau T_u \xi du, \qquad k = 0, 1, 2, \dots$$

is stationary and $\mathbf{P}|\xi_0| \le \mathbf{P} \int_0^\tau T_u |\xi| du < \infty$. It follows from Corollary 7 that;

$$\lim_{n \to \infty} \frac{1}{n} \int_0^{\tau_n} T_u \xi du = \mathbf{P}\left(\left(\int_0^\tau T_u \xi | \mathcal{T}^\bullet \right) \right) \tag{24}$$

is almost certainly valid, where the σ-algebra $\mathcal{T}^\bullet$ consists of invariant (with respect to T_τ), measurable (with respect to $\mathcal{M}$) events. This implies that the sequence τ_n, $n = 1, 2, \dots$, is almost certainly unbounded which, together with the inequality;

$$\frac{1}{n} \int_0^{\tau_n} T_u \xi du \le \frac{1}{n} \int_0^t T_u \xi du \le \frac{1}{n} \int_0^{\tau_{n+1}} T_u \xi du,$$

that is valid for $\xi \ge 0$ and $\tau_n \le t < \tau_{n+1}$, implies that the limit given in (23) exists with unit probability and is equal to $\mathbf{P}\left(\int_0^\tau T_u \xi du | \mathcal{T}^\bullet\right) / \mathbf{P}\left(\int_0^\tau T_u \eta du | \mathcal{T}^\bullet\right)$. Hence, if the left hand side of (23) is denoted by ζ, then $\mathbf{P}\left(\zeta \int_0^\tau T_u \eta du\right) = \mathbf{P} \int_0^\tau T_u \xi du$. But, because (24) implies the divergence of the integral $\int_0^\infty T_u \eta du$ with unit probability for $\eta > 0$, the variable ζ is measurable with respect to the σ-algebra $\mathcal{T}$, which is a σ-subalgebra of $\mathcal{T}^\bullet$. This implies that ζ is almost certainly equal to the right hand side of (23).

14. Starting with an arbitrary random translation operator T_τ, which preserves the probability $\mathbf{P}$ on $\mathcal{M}$, a number of other translation operators can be constructed, which preserve generally other measures on $\mathcal{M}$ or on a subset of $\mathcal{M}$.

In this way, if a measurable, with respect to $\mathcal{M}$, variable ν takes the values $\{1, 2, \dots, \infty\}$, $\mathbf{P}(\nu < \infty) > 0$ and:

$$\{\nu > 1\} \subset \{T_\tau \nu = \nu - 1\}, \tag{25}$$

then the translation operator T_{τ_ν} is such that;

$$T_{\tau_\nu} \xi = \begin{cases} T_\tau^k \xi & \text{if} \quad \nu = k < \infty, \\ 0 & \text{if} \quad \nu = \infty, \end{cases}$$

preserves the probability $\mathbf{Q}$ on $\mathcal{M}$;

$$\mathbf{Q}\xi = \frac{1}{\mathbf{P}(\nu = 1)} \mathbf{P}(\nu = 1, T_\tau \xi).$$

Indeed;

$$\mathbf{P}(\nu = 1, T_\tau\xi) = \mathbf{P}\xi - \mathbf{P}(\nu > 1, T_\tau\xi),$$

and if $\nu > 1$, then $T_\tau T_{\tau_\nu}\xi = T_{\tau_\nu}\xi$. Thus

$$\begin{aligned}\mathbf{P}(\nu = 1,\, T_\tau^{-1}T_{\tau_\nu}^{-1}\Gamma) &= \mathbf{P}(T_{\tau_\nu}^{-1}\Gamma) - \mathbf{P}(\nu > 1,\, (T_{\tau_\nu}T_\tau)^{-1}\Gamma)\\ &= \mathbf{P}(T_{\tau_\nu}^{-1}\Gamma) - \mathbf{P}(\nu > 1,\, T_{\tau_\nu}^{-1}\Gamma) = \mathbf{P}(\nu = 1, T_{\tau_\nu}^{-1}\Gamma)\\ &= \mathbf{P}(\nu = 1,\, T_\tau^{-1}\Gamma).\end{aligned}$$

Proof that the probability $\mathbf{P}(\nu = 1)$ is positive follows from (25) because, if $\mathbf{P}(\nu > 1) = 1$, then

$$\begin{aligned}1 = \mathbf{P}(T_\tau\nu > 1,\, \nu > 1) &= \mathbf{P}(\nu > 2) = \mathbf{P}(T_\tau\nu > 2,\, \nu > 1)\\ &= \mathbf{P}(\nu > 3) = \ldots = \mathbf{P}(\nu > k) = \ldots,\end{aligned}$$

which contradicts the fact that ν is finite with a positive probability.

On the set $\{\nu < \infty\}$, the probability $\mathbf{P}$ can be expressed in terms of $\mathbf{Q}$ by the formula:

$$\frac{1}{\mathbf{P}(\nu < \infty)}\mathbf{P}(\xi, \nu < \infty) = \frac{1}{\mathbf{Q}(\nu, \nu < \infty)}\mathbf{Q}\left(\sum_{k=0}^{\nu-1} T_\tau^k\xi,\ \nu < \infty\right), \tag{26}$$

which is a discrete analogue of formula (22). It follows from (22) that the mean $\mathbf{Q}(\nu,\, \nu < \infty)$ is finite:

$$\begin{aligned}\mathbf{P}(\nu = 1)\mathbf{Q}(k < \nu < \infty) &= \mathbf{P}(\nu = 1,\, k < T_\tau\nu < \infty)\\ &= \mathbf{P}(k < \nu < \infty) - \mathbf{P}(\nu > 1,\, k < T_\tau\nu < \infty)\\ &\mathbf{P}(k < \nu < \infty) - \mathbf{P}(k = 1 < \nu < \infty) = \mathbf{P}(\nu = k = 1);\\ \mathbf{Q}(\nu, \nu < \infty) &= \sum_{k=0}^{\infty}\mathbf{Q}(\nu > k,\, \nu < \infty) = \frac{\mathbf{P}(\nu < \infty)}{\mathbf{P}(\nu = 1)}.\end{aligned}$$

Formula (26) can also be demonstrated by the following:

$$\begin{aligned}&\mathbf{P}(\nu = 1)\mathbf{Q}\left(\sum_{k=0}^{\nu-1} T_\tau^k\xi,\ \nu < \infty\right)\\ &= \mathbf{P}\left(\nu = 1,\ T_\tau\sum_{k=0}^{\nu-1} T_\tau^k\xi,\ T_\tau\nu < \infty\right)\\ &= \mathbf{P}\left(\sum_{k=0}^{\nu-1} T_\tau^k\xi,\ \nu < \infty\right) - \mathbf{P}\left(\infty > \nu > 1,\ \sum_{k=0}^{T_\tau\nu-1} T_\tau^{k+1}\xi\right)\\ &= \mathbf{P}\left(\sum_{k=0}^{\nu-1} T_\tau^k\xi,\ \nu < \infty\right) - \mathbf{P}\left(\infty > \nu > 1,\ \sum_{k=1}^{\tau-2} T_\tau^{k+1}\xi\right)\\ &= \mathbf{P}\left(\sum_{k=0}^{\nu-1} T_\tau^k\xi,\ \nu < \infty\right) - \mathbf{P}\left(1 < \nu < \infty,\ \sum_{k=1}^{\nu-1} T_\tau^k\xi\right)\\ &= \mathbf{P}(\xi,\ \nu < \infty).\end{aligned}$$

15. Let now $\Delta \in \mathcal{M}$, $\mathbf{P}(\Delta) > 0$, and:

$$\nu = \nu(\omega) = \min\{k \geq 1 : \omega \in T_\tau^{-k}\Delta\}. \tag{27}$$

Then, $\mathbf{P}(\nu < \infty) > 0$ because otherwise the series $\sum_{k=1}^{\infty} T_\tau^k i_\Delta$ converges with unit probability, which contradicts Corollary 7;

$$\mathbf{P}\left[\lim_{n\to\infty} \frac{1}{n}\sum_{k=0}^{n-1} T_\tau^k i_\Delta\right] = \mathbf{P}(\Delta).$$

For such a choice of ν, condition (25) is valid and the probability $\mathbf{Q}$ is the conditional probability $\mathbf{P}$ with the condition Δ. Indeed;

$$\mathbf{P}(\nu = 1)\mathbf{Q}(\Gamma) = \mathbf{P}(\nu = 1), T_\tau^{-1}\Gamma) = \mathbf{P}(T_\tau^{-1}\Delta, T_\tau^{-1}\Gamma) = \mathbf{P}(\Delta\Gamma)$$

and:

$$\mathbf{P}(\nu = 1) = \mathbf{P}(T_\tau^{-1}\Delta) = \mathbf{P}(\Delta)$$

so that;

$$\mathbf{Q}(\Gamma) = \mathbf{P}(\Gamma|\Delta) = \mathbf{P}_\Delta(\Gamma).$$

If a variable ν is defined by (27), then the transformation T_{τ_ν} preserves the conditional probability $\mathbf{P}_\Delta$. If the transformation T_τ is ergodic, then:

$$\lim_{n\to\infty} \frac{1}{n}\sum_{k=0}^{n-1} T_\tau^k i_\Delta = \mathbf{P}(\Delta)$$

is almost certainly true and, consequently, $\mathbf{P}(\nu < \infty) = 1$. In this case, formula (26) gives an expression for the probability $\mathbf{P}$ in terms of $\mathbf{P}_\Delta$:

$$\mathbf{P}\xi = \frac{1}{\mathbf{P}_\Delta \nu}\mathbf{P}_\Delta \sum_{k=0}^{\nu-1} T_\tau^k \xi,$$

where the variable ξ is bounded and is measurable with respect to $\mathcal{M}$. This implies the following statement.

Proposition. *The probability* $\mathbf{P}$ *on the* σ*-algebra* $\mathcal{M}$*, which is generated by a stationary ergodic process, can be uniquely recovered from any of the conditional probabilities* $\mathbf{P}_\Delta$ *where* $\Delta \in \mathcal{M}$, $\mathbf{P}(\Delta) > 0$.

16. The requirement that the measure $\mathbf{P}$ is a probability measure is not included and is replaced with the assumption that the measure is finite with respect to σ.

Proposition.

Let the transformation T_τ preserve a finite (with respect to σ) measure $\mathbf{P}$ on $\mathcal{M}$. Then, if $\Delta \in \mathcal{M}$, $0 < \mathbf{P}(\Delta) < \infty$, and $\mathbf{P}\left(\rceil \cup_{k=1}^{\infty} T_{\tau_k}^{-1}\Delta\right) = 0$, then the transformation T_{τ_ν}, where ν is defined by (27), preserves the probability measure $\mathbf{P}_\Delta(\Gamma) = \mathbf{P}(\Gamma\Delta)/\mathbf{P}(\Delta)$, where $\Gamma \in \mathcal{M}$, and $\mathbf{P}\xi = \mathbf{P}(\Delta)\mathbf{P}_\Delta \sum_{k=0}^{\nu-1} T_\tau^k \xi$.

17. The transformations of random processes which preserve the conditions of Theorems 12 and 13 are considered. One of the most widely used transformations in the theory of random processes is a random change of variables, which is now considered.

Let the space of trajectories of a process $\mathfrak{z}_t$ $(t \geq 0)$ be closed with respect to any continuously increasing unbounded change of time. This implies that, for any $\omega \in \Omega$ and any continuously increasing unbounded function $u(t)$ ($t \geq 0$ and $u(0) = 0$) there exists an $\omega' \in \Omega$ such that $\mathfrak{z}_{u(t)}(\omega) = \mathfrak{z}_t(\omega')$ for all $t \geq 0$.

Then, if a measurable (with respect to $\mathcal{M}$) random process $\mathfrak{F}(t)$ $(t \geq 0)$ continuously increases from zero to infinity, then the space of trajectories of the process $\mathfrak{h}_t = \mathfrak{z}_{\mathfrak{t}(t)}$ $(t \geq 0)$, where $\mathfrak{F}(\mathfrak{t}(t)) = \mathfrak{t}(\mathfrak{F}(t)) = t$, is stable with respect to the translations. Consequently, the process $\mathfrak{h}_t$ $(t \geq 0)$ is associated with a family of translation operators S_u $(u \geq 0)$ and with the σ-algebra $\mathcal{J}$ of events, invariant with respect to S_u $(u \geq 0)$, from the σ-algebra $\mathfrak{G}$ generated by the process $\mathfrak{h}_t$ $(t \geq 0)$.

18. Theorem. *Let the translation operator T_τ preserve the probability $\mathbf{P}$ on $\mathcal{M}$ and $\mathbf{P}(0 < \tau < \infty) = 1$. Then, if:*

$$\mathfrak{F}(\tau + t) = \mathfrak{F}(\tau) + T_\tau \mathfrak{F}(t) \tag{28}$$

for all $t \geq 0$ and $\omega \in \Omega$, then:

$$\lim_{t\to\infty} \frac{\int_0^t S_u \xi du}{\int_0^t S_u \eta du} = \frac{\mathbf{P}\left[\int_0^{\mathfrak{F}(\tau)} S_u \xi du | \mathcal{J}\right]}{\mathbf{P}\left[\int_0^{\mathfrak{F}(\tau)} S_u \eta du | \mathcal{J}\right]} \tag{29}$$

is almost certainly true, where $\mathbf{P} \int_0^{\mathfrak{F}(\tau)} S_u(|\xi| + |\eta|) du < \infty$, $\mathbf{P}\left[\int_0^{\mathfrak{F}(\tau)} S_u \eta du | \mathcal{J}\right] \neq 0$ is almost certainly valid.

Proof. The σ-algebra $\mathfrak{G}$ generated by the process $\mathfrak{h}_t$ $(t \geq 0)$ is a subalgebra of the σ-algebra $\mathcal{M}$ so that the translation operators T_t $(t \geq 0)$ can be applied to measurable events with respect to $\mathfrak{G}$. It follows from (28) that the translation operators $S_{\mathfrak{F}(\tau)}$ and T_τ coincide with measurable events with respect to $\mathfrak{G}$. Indeed;

$$S_{\mathfrak{F}(\tau)} \mathfrak{h}_t = \mathfrak{h}_{t+\mathfrak{F}(\tau)} = \mathfrak{z}_{\mathfrak{t}+\mathfrak{F}(\tau)},$$

$$T_\tau \mathfrak{h}_t = T_\tau \mathfrak{z}_{\mathfrak{t}(t)} = \mathfrak{z}_{\tau + T_\tau \mathfrak{t}(t)}.$$

However, as equation (28) demonstrates;

$$\begin{aligned} \tau + T_\tau \mathfrak{t}(t) &= \mathfrak{t}(\mathfrak{F}(\tau + T_\tau \mathfrak{t}(t))) = \mathfrak{t}(\mathfrak{F}(\tau) + T_\tau \mathfrak{F}(\mathfrak{t}(t))) \\ &= \mathfrak{t}(\mathfrak{F}(\tau) + t), \end{aligned}$$

and consequently, $S_{\mathfrak{G}(\tau)} \mathfrak{h}_t = T_\tau \mathfrak{h}_t$. This proves that the operators $S_{\mathfrak{G}(\tau)}$ and T_τ coincide with measurable variables with respect to $\mathfrak{G}$.

If the variable $\mathfrak{F}(\tau)$ were measurable with respect to $\mathfrak{G}$, then the proof of the theorem follows directly from Theorem 4.

Although the variable $\mathfrak{F}(\tau)$ is not, in general, measurable with respect to $\mathfrak{G}$, the proof of Theorem 4 can be modified so that it can be applied to this case. This modification is based on the fact that the operators $S_{\mathfrak{F}(\tau_k)}$ and T_{τ_k} coincide with measurable, with respect to $\mathfrak{G}$, variables. This follows from the following equation:

$$\mathfrak{F}(\tau_k + t) = \mathfrak{F}(\tau_k) + T_{\tau_k}\mathfrak{F}(t),$$

a proof of which follows from (28) if the substitution $t = u + T_\tau \tau_{k-1}(\omega)$ is made. Then;

$$\begin{aligned} \mathfrak{F}(\tau_k(\omega) + u, \omega) &= \mathfrak{F}(\tau(\omega), \omega) + \mathfrak{F}(u + T_\tau \tau_{k-1}(\omega), T_\tau \omega) \\ &= \mathfrak{F}(\tau(\omega), \omega) + \mathfrak{F}(u + \tau_{k-1}(T_\tau \omega), T_\tau \omega), \end{aligned}$$

i.e. $\mathfrak{F}(\tau_k + u) = \mathfrak{F}(\tau) + T_\tau \mathfrak{F}(\tau_{k-1} + u)$. Iterating this formula, the proof is obtained.

It can be shown that the sequence $\int_{\mathfrak{F}(\tau_k)}^{\mathfrak{F}(\tau_{k+1})} S_u \xi du$, $k = 0, 1, 2, \ldots$, is stationary. This follows from the following analysis:

$$\begin{aligned} T_\tau \int_{\mathfrak{F}(\tau_{k-1})}^{\mathfrak{F}(\tau_k)} S_u \xi du &= \int_{T_\tau \mathfrak{F}(\tau_{k-1})}^{T_\tau \mathfrak{F}(\tau_k)} T_\tau S_u \xi du = \int_{\mathfrak{F}(\tau_k) - \mathfrak{F}(\tau_{k-1})}^{\mathfrak{F}(\tau_{k-1}) - \mathfrak{F}(\tau_k)} S_{\mathfrak{F}(\tau) + u} \xi du \\ &= \int_{\mathfrak{F}(\tau_k)}^{\mathfrak{F}(\tau_{k+1})} S_u \xi du. \end{aligned}$$

From Corollary 7;

$$\lim_{n \to \infty} \frac{1}{n} \int_0^{\mathfrak{F}(\tau_n)} S_u \xi du = \mathbf{P}\left(\int_0^{\mathfrak{F}(\tau)} S_u \xi | \mathcal{I}^\bullet \right)$$

is almost certainly true, where the σ-algebra $\mathcal{I}^\bullet$ consists of invariant (with respect to T_τ) measurable (with respect to $\mathcal{M}$) events. The conclusion of the proof is the same as has already been given for Theorem 4.

19. An example of a function $\mathfrak{F}(t)$ $(t \geq 0)$ which satisfies condition (28), is the additive functional of integral type, for example;

$$\mathfrak{F}(t) = \int_0^t T_u \zeta du, \tag{30}$$

where the variable ζ is positive and measurable with respect to $\mathcal{M}$. In this case, $\mathfrak{F}(s + t) = \mathfrak{F}(s) + T_s \mathfrak{F}(t)$ for all $s, t \geq 0$, and the right hand side of (29) becomes:

$$\frac{\mathbf{P}\left[\int_0^\tau T_u(\xi\eta) du | \mathcal{I}\mathfrak{G}\right]}{\mathbf{P}\left[\int_0^\tau T_u(\eta\zeta) du | \mathcal{I}\mathfrak{G}\right]}. \tag{31}$$

The operators T_u and $S_{\mathfrak{F}(u)}$ coincide with measurable (with respect to $\mathfrak{G}$) variables. So the σ-algebra $\mathfrak{J}$ is equal to the intersection $\mathcal{I}\mathfrak{G}$ of the σ-algebras $\mathcal{I}$ and $\mathfrak{G}$. Moreover, if $\xi = \varphi(\mathfrak{h}_{s_1}, \ldots, \mathfrak{h}_{s_n})$, then:

$$S_u\xi = \varphi(\mathfrak{h}_{u+s_1}, \ldots, \mathfrak{h}u + s_n) = \varphi(\mathfrak{z}_{\mathfrak{F}^{-1}(u+s_1)}, \ldots, \mathfrak{z}_{\mathfrak{F}^{-1}(u+s_n)})$$

and, after the change of variable $u \to \mathfrak{F}(u)$;

$$\begin{aligned} \int_0^{\mathfrak{F}(\tau)} S_u\xi du &= \int_0^{\tau} \varphi(\mathfrak{z}_{\mathfrak{F}^{-1}(\mathfrak{F}(u)+s_1)}, \ldots, \mathfrak{z}_{\mathfrak{F}^{-1}(\mathfrak{F}(u)+s_n)})T_u\zeta du \\ &= \int_0^{\tau} \varphi(\mathfrak{z}_{u+T_u\mathfrak{F}^{-1}(s_1)}, \ldots, \mathfrak{z}_{u+T_u\mathfrak{F}^{-1}(s_n)})T_u\zeta du \\ &= \int_0^{\tau} T_u[\varphi(\mathfrak{z}_{\mathfrak{F}^{-1}(s_1)}, \ldots, \mathfrak{z}_{\mathfrak{F}^{-1}(s_n)})\zeta] \\ &= \int_0^{\tau} T_u(\xi\zeta)du. \end{aligned}$$

20. If the process $\mathfrak{z}_t$ $(t \geq 0)$ is stationary, further simplifications can be made by setting $\tau = 1$. Then (31) becomes equal to $\mathbf{P}(\xi\zeta|\mathfrak{G}\mathcal{I})/\mathbf{P}(\eta\xi|\mathcal{I}\mathfrak{G})$. Consequently, if the process $\mathfrak{z}_t$ $(t \geq 0)$ is stationary and $\mathfrak{h}_t = \mathfrak{z}_{\mathfrak{F}^{-1}(t)}$, where $\mathfrak{F}(t)$ is defined by formula (30), in which $\mathbf{P}\zeta < \infty$, then (in the case where $\mathbf{P}|\xi\zeta| < \infty$);

$$\lim_{t\to\infty} \frac{1}{t} \int_0^t S_u\xi du = \frac{\mathbf{P}(\xi\zeta|\mathcal{I}\mathfrak{G})}{\mathbf{P}(\zeta|\mathcal{I}\mathfrak{G})} \tag{32}$$

is almost certainly true.

The right hand side is equal to $\mathbf{P}^\bullet(\xi|\mathcal{I}\mathfrak{G})$. The probability $\mathbf{P}^\bullet$ is obtained from the probability $\mathbf{P}$ by using the absolute continuous transformation with the density $\zeta^\bullet = \frac{\zeta}{\mathbf{P}(\zeta|\mathcal{I}\mathfrak{G})}$, i.e. $\mathbf{P}^\bullet\eta = \mathbf{P}\eta\zeta^\bullet$. The right hand side does not change if ξ is replaced with $S_t\xi$ for any $t > 0$. This means that the probability $P^\bullet$ is invariant with respect to S_t. In other words, the process $\mathfrak{h}_t$ $(t \geq 0)$ is stationary on the space $(\Omega, \mathcal{M}, \mathbf{P}^\bullet)$. This fact is important and is made in a separate statement.

21. Proposition. *The change of time $t \to \mathfrak{F}^{-1}(t)$, where $\mathfrak{F}(t) = \int_0^t T_u\zeta du$, $\zeta > 0$ and $\mathbf{P}\zeta < \infty$, together with an absolutely continuous change of probability measure $\mathbf{P}$ (with density $\zeta/\mathbf{P}(\zeta|\mathcal{I})$, transforms a stationary process into another stationary process.*

22. The random change of time is a particular example of a general transformation of the space, Ω, of elementary events. Let $\mathfrak{h}_t(\omega) = \mathfrak{z}_t(U\omega)$, where $\omega \to U\omega$ is a measurable mapping of the measurable space $(\Omega, \mathcal{M})$ into itself. Then, if the process $\mathfrak{z}_t$ $(t \geq 0)$ satisfies the conditions of Theorem 4 and:

$$T_\tau U\omega = UT_\tau\omega \tag{33}$$

for all $\omega \in \Omega$, then the transformation S_σ of the translation of a trajectory of the process $\mathfrak{h}_t$ $(t \geq 0)$ by the value $\sigma(\omega) = \tau(U\omega)$ preserves the probability $\mathbf{P}$ on the σ-algebra $\mathfrak{G}$, generated by the process $\mathfrak{h}_t$ $(t \geq 0)$. Indeed, $S_\sigma \mathfrak{h}_t(\omega) = \mathfrak{z}_{\tau(U)+t}(U\omega) = \mathfrak{z}_t(T_\tau U\omega)$ and $T_\tau \mathfrak{h}_t(\omega) = \mathfrak{h}_t(UT_\tau\omega)$. Consequently, the operators S_σ and T_τ coincide on measurable (with respect to $\mathfrak{G}$) variables and so the probability $\mathbf{P}$ on $\mathfrak{G}$ is invariant with respect to S_σ.

If the random change of time $t \to \mathfrak{t}(t,\omega)$ is considered as a transformation U of the space of elementary events, i.e. $\mathfrak{z}_t(U\omega) = \mathfrak{z}_{\mathfrak{t}(t,\omega)}(\omega)$, then, for condition (33) to be valid, it is necessary, together with condition (22), to impose the following additional condition:

$$\tau(U\omega) = \mathfrak{F}(\tau(\omega), \omega). \tag{34}$$

Indeed, $\mathfrak{z}_t(UT_\tau\omega) = \mathfrak{z}_{\mathfrak{t}(\tau(U\omega)+t,\omega)}(\omega)$ and $\mathfrak{z}_t(T_\tau U\omega\omega) = \mathfrak{z}_{\tau(U\omega)+t,\omega)}(\omega)$. This implies that, as long as the structure of the trajectories of the process, $\mathfrak{z}_t$ $(t \geq 0)$, is not considered, it is necessary to postulate either of the following conditions:

$$\mathfrak{t}(\tau(U\omega) + t, \omega) = \tau(\omega)) + \mathfrak{t}(t, T_\tau\omega),$$

or;

$$\tau(U\omega) + t = \mathfrak{F}(\tau(\omega) + \mathfrak{t}(t, T_\tau\omega), \omega), \tag{35}$$

which becomes condition (34) if $t = 0$. If condition (34) is valid, then relation (35) becomes $\mathfrak{F}(\tau(\omega)+u,\omega) = \mathfrak{F}(\tau(\tau(\omega),\omega)+\mathfrak{F}(u,T_\tau\omega)$ if $u = \mathfrak{t}(t,T_\tau\omega)$ and $t = \mathfrak{F}(t,T_\tau\omega)$. In this case the value of t in condition (35) is arbitrary because u is arbitrary.

So to make condition (33) valid, it is required that condition (34) and condition (28) are valid.

The condition $\tau(U\omega) = \mathfrak{F}(\tau(\omega),\omega)$ is valid as an identity if, for example, τ is the time at which the process $\mathfrak{z}_t$ $(t \geq 0)$ reaches the set of states $A \in \mathcal{A}$ for the first time, i.e. $\tau = \tau_A = \inf\{t \geq 0 : \mathfrak{z}_t \in A\}$. The same is true if τ is the time at which the process returns to A through B, i.e. $\tau = \tau_B + T_{\tau_B}\tau_A$, etc.

At the same time, this condition will not be valid if the time τ is not random and if $\mathfrak{F}(t)$ is given by (30) with a nontrivial variable ζ.

1.3 Examples

23. *A sequence of independent identically distributed variables* $\mathfrak{z}_0, \mathfrak{z}_1, \mathfrak{z}_2, \dots$, which take values in a phase space $(E, \mathcal{A})$ gives the simplest example of a stationary ergodic sequence. It is clear that it is stationary, and Kolmogorov's law of zero and identity implies that the σ-algebra $\mathcal{I}$ of invariant events is trivial. This situation is so simple that a direct proof can be given.

Let a measurable (with respect to $\mathcal{I}$) variable η be bounded. Then, by Levi's theorem;

$$\eta = \lim_{n\to\infty} \mathbf{P}(\eta|\mathfrak{z}_0, \mathfrak{z}_1, \dots, \mathfrak{z}_n) = \lim_{n\to\infty} \mathbf{P}(T^{n+1}\eta|\mathfrak{z}_0, \mathfrak{z}_1, \dots, \mathfrak{z}_n)$$

is almost certainly true, where T is the translation operator of the sequence $\mathfrak{z}_n$, $n = 0, 1, 2, \dots$, by unity. The variable $T^{n+1}\eta$ is measurable with respect to $\mathfrak{z}_{n+1}$, $\mathfrak{z}_{n+1}$, ... and, consequently, it does not depend on the variables $\mathfrak{z}_0$, $\mathfrak{z}_1$, ... , $\mathfrak{z}_n$. Hence;

$$\eta = \lim_{n\to\infty} \mathbf{P}(T^{n+1}\eta) = \mathbf{P}\eta$$

is almost certainly valid.

24. *Permutable* variables give a more interesting example.

A sequence $\mathfrak{z}_0, \mathfrak{z}_1, \dots$ with values in $(E, \mathcal{A})$ is called permutable if, for any natural n, all $n!$ vectors $(\mathfrak{z}_{i_1}, \mathfrak{z}_{i_2}, \dots, \mathfrak{z}_{i_n})$, $i_1, i_2, \dots, i_n = 0, 1, \dots, n-1$, have an identical distribution.

A permutable sequence is automatically stationary. Indeed;

$$\begin{aligned}
&\mathbf{P}(\mathfrak{z}_0 \in A_0, \mathfrak{z}_1 \in A_1, \dots, \mathfrak{z}_{n-1} \in A_{n-1}) = \\
&\quad = \mathbf{P}(\mathfrak{z}_0 \in A_0, \mathfrak{z}_1 \in A_1, \dots, \mathfrak{z}_{n-1} \in A_{n-1}, \mathfrak{z}_n \in E) \\
&\quad = \mathbf{P}(\mathfrak{z}_0^1 \in A_0, \mathfrak{z}_1^2 \in A_1, \dots, \mathfrak{z}_n \in A_{n-1}, \mathfrak{z}_0 \in E) \\
&\quad = \mathbf{P}(\mathfrak{z}_0 \in A_0, \mathfrak{z}_1 \in A_1, \dots, \mathfrak{z}_n \in A_{n-1}).
\end{aligned}$$

A permutable sequence is ergodic (implying that its σ-algebra $\mathcal{I}$ of invariant events is trivial) if and only if it degenerates into a sequence of independent variables. This follows from the conditional indepedence of $\mathfrak{z}_0$, $\mathfrak{z}_1$, ... , $\mathfrak{z}_n$ under the condition $\mathcal{I}$, i.e.

$$\mathbf{P}(\mathfrak{z}_0 \in A, \mathfrak{z}_1 \in A_1, \dots, \mathfrak{z}_m \in A_m|\mathcal{I}) = \prod (\mathfrak{z}_k \in A_k|\mathcal{I})$$

is almost certainly true. This equation is equivalent to the following:

$$\mathbf{P}(\eta f_1(\mathfrak{z}_1) \dots f_m(\mathfrak{z}_m)) = \mathbf{P}(\eta \zeta_1 \dots \zeta_m), \tag{36}$$

where the variable η is bounded and is measurable with respect to $\mathcal{I}$. The functions f_i are bounded and are measurable with respect to $\mathcal{A}$ and $\zeta_i = \mathbf{P}(f_i(\mathfrak{z}_i)|\mathcal{I}) = \mathbf{P}(f_i(\mathfrak{z}_0)|\mathcal{I})$, $i = 1, \dots, m$.

The right hand side of (36) can also be written as:

$$\begin{aligned}
&\lim_{n\to\infty} n^{-m} \mathbf{P}\left(\eta \sum_{k=1}^{n} f_1(\mathfrak{z}_k) \dots \sum_{k=1}^{n} f_m(\mathfrak{z}_k)\right) = \\
&\quad \lim_{n\to\infty} n^{-m} \mathbf{P}\left[\left(T^{n+1}\eta \sum_{k_1,\dots,k_m=1}^{n} f_1(\mathfrak{z}_{k_1}) \dots f_m(\mathfrak{z}_{k_m})\right)\right].
\end{aligned} \tag{37}$$

The variable $T^{n+1}\eta$ is measurable with respect to $\mathfrak{z}_{n+1}, \mathfrak{z}_{n+2}, \dots$. Therefore, because of the property of exchangeability;

$$\mathbf{P}[(T^{n+1}\eta)f_1(\mathfrak{z}_{k_1})\dots f_m(\mathfrak{z}_{k_m})] = \mathbf{P}[(T^{n+1}\eta)f_1(\mathfrak{z}_1)\dots f_m(\mathfrak{z}_m)],$$

is true if there are no equal indices among $k_1, \dots, k_m$. But the number of such collections of indices $(k_1, \dots, k_m)$ is equal to $n(n-1)\dots(n-m+1)$ and, as n tends to ∞, it asymptotically equals n^n. Consequently limit (37) and, hence, the right hand side of (36) coincide with:

$$\mathbf{P}[(T^{n+1}\eta)f_1(\mathfrak{z}_1)\dots f_m(\mathfrak{z}_m)] = \mathbf{P}[\eta f_1(\mathfrak{z}_1)\dots f_m(\mathfrak{z}_m)].$$

25. In this example a *random walk on a circle with unit circumference* is considered. The position ζ_n of a wandering particle after the nth step is given by the length of the arc from some fixed point measured in a fixed direction, either clockwise or anticlockwise. Subsequent displacements are independent and identically distributed. Hence;

$$\zeta_n = \zeta_0 + \xi_1 + \dots + \xi_n (\operatorname{mod} 1), \quad \zeta \in [0,1),$$

where the variables $\xi_1, \xi_2, \dots$ are independent. These do not depend on the initial position ζ_0, and have the same distribution F, concentrated on $[0,1)$. The equation;

$$\begin{aligned}\mathbf{P}_x(\zeta_1 < c_1, \zeta_2 < c_2, \dots, \zeta_n < c_n) = \mathbf{P}(c_1 > x + \xi_1 (\operatorname{mod} 1),\\ c_2 > x + \xi_1 + \xi_2 (\operatorname{mod} 1), \dots,\\ c_n > x + \xi_1 + \dots + \xi_n (\operatorname{mod} 1)),\end{aligned}$$

where $x, c_1, c_2, \dots$ $(c_n \in [0,1))$, defines, on the σ-algebra $\mathcal{N}$ generated by the sequence $\zeta_0, \xi_1, \dots$, a family of probability measures $\mathbf{P}_x$, $(x \in [0,1))$, which will be a regular variant of the conditional probabilities if $\zeta_0 = x$.

The sequence $\zeta_0, \zeta_1, \dots$ is stationary with the probability $\mathbf{P}^\bullet = \int_0^1 dx \mathbf{P}_x$. Indeed;

$$\begin{aligned}\mathbf{P}^\bullet(\zeta_1 < c) &= \mathbf{P}^\bullet(\zeta_0 + \xi_1 < c, \zeta_0 + \xi_1 < 1)\\ &+ \mathbf{P}^\bullet(\zeta_0 + \xi_1 - 1 < c, \zeta_0 + \xi_1 \geq 1)\\ &= \mathbf{P}^\bullet(\zeta_0 + \xi_1 < c) + \mathbf{P}^\bullet(1 \leq \zeta_0 + \xi_1 < 1 + c)\\ &= \int_0^c \mathbf{P}(\xi_1 < c - x)dx + \int_c^1 \mathbf{P}(1 - x \leq \xi_1 < 1 + c + x)dx\\ &+ \int_0^c \mathbf{P}(\xi_1 \geq 1 - x)dx = c\end{aligned}$$

if $0 \leq c < 1$, i.e. if the distribution of ζ_0 is uniform, then ζ_1 is also distributed uniformly. Consequently;

$$\mathbf{P}^\bullet(\zeta_0 < c_0, \zeta_1 < c_1, \dots, \zeta_n < c_n) =$$

$$= \int_0^\infty \mathbf{P}(c_1 > \xi_1 + x(\mathrm{mod}\,1), c_2 > \xi_2 + \xi_1$$
$$+ x(\mathrm{mod}\,1), \dots, c_n > \xi_n + \dots + \xi_1 + x(\mathrm{mod}\,1))dx$$
$$= \int_0^\infty \mathbf{P}(c_1 > \xi_2 + x(\mathrm{mod}\,1), c_2 > \xi_3 + \xi_2$$
$$+ x(\mathrm{mod}\,1), \dots, c_n > \xi_{n+1} + \dots + \xi_2 + x(\mathrm{mod}\,1))dx$$
$$= \mathbf{P}^\bullet c(\xi_1 < c_0, c_1 > \xi_2 + \zeta_1(\mathrm{mod}\,1), c_2 > \xi_3 + \xi_2$$
$$+ \zeta_1(\mathrm{mod}\,1), \dots, c_n > \xi_{n+1} + \dots + \xi_2 + \zeta_1(\mathrm{mod}\,1))$$
$$= \mathbf{P}^\bullet(\xi_1 < c_0, \zeta_2 < c_1, \zeta_3 < c_3, \dots, \zeta_{n+1} < c_n).$$

If the distribution, F, of the variable ξ_1 is not concentrated at the points 0, $1/m$, $2/m, \dots, (m-1)/m$ for any natural m, then the sequence is ergodic with respect to the measure $\mathbf{P}^\bullet$. A proof of this is based on the *Markovian property* of the sequence ζ_n, $n = 0, 1, 2, \dots$. Namely, if a random variable, η, is bounded and is measurable with respect to $\mathcal{N}$, then:

$$\mathbf{P}_x(T^n \eta | \zeta_1, \dots, \zeta_n) = \varphi(\zeta_n) \tag{38}$$

is almost certainly valid with respect to $\mathbf{P}_x$, where $\varphi(y) = \mathbf{P}_y \eta$, and T is the unit translation operator of the sequence ζ_n, $n = 0, 1, 2, \dots$.

Indeed, if $(\zeta_1, \dots, \zeta_n)$ does not depend on $(\xi_{n+1}, \dots, \xi_{n+m})$ then:

$$\mathbf{P}_x(\zeta_1 \in A_1, \dots, \zeta_n \in A_n, \zeta_{n+1} \in B_1, \dots, \zeta_{n+m} \in B_m) =$$
$$= \mathbf{P}_x\{\zeta_1 \in A_1, \dots, \zeta_n \in A_n, \zeta_n + \xi_{n=1}(\mathrm{mod}\,1) \in B_1, \dots, \zeta_n$$
$$+ \xi_{n+1} + \dots + \xi_{n+m}(\mathrm{mod}\,1) \in B_m\}$$
$$= \int_{A_n} \mathbf{P}_x\{\zeta_1 \in A_1, \dots, \zeta_{n-1} \in A_{n-1}, \zeta_n \in dy, y$$
$$+ \xi_{n+1}(\mathrm{mod}\,1) \in B_1, \dots, y + \xi_{n+1} + \dots$$
$$+ \xi_{n+m}(\mathrm{mod}\,1) \in B_m\}$$
$$= \int_{A_n} \mathbf{P}_x(\zeta_1 \in A_1, \dots, \zeta_{n-1} \in A_{n-1}, \zeta_n \in dy$$
$$\times \mathbf{P}(y + \xi_1(\mathrm{mod}\,1) \in B_1, \dots, y + \xi_1 + \dots + \xi_m(\mathrm{mod}\,1) \in B_m)$$
$$= \int_{A_n} \mathbf{P}_x(\zeta \in A_1, \dots, \zeta_{n-1} \in A_{n-1}, \zeta_n \in dy) \mathbf{P}_y(\zeta_1 \in B_1, \dots, \zeta_m \in B_m)$$
$$= \mathbf{P}_x(\zeta_1 \in A_1, \dots, \zeta_n \in A_n, \psi(\zeta_n)),$$

where $\psi(y) = \mathbf{P}_y(\zeta_1 \in B_1, \dots, \zeta_m \in B_m)$. $A_1, \dots < A_n$, $B_1, \dots, B_m$ are Borel subsets of $[0, 1)$. The equation obtained is equivalent to (38).

Consequently, if a measurable (with respect to $\mathcal{N}$) variable, η, is bounded, invariant (with respect to T) and $\varphi(y) = \mathbf{P}_y \eta$, then:

$$\eta \;=\; \lim_{n \to \infty} \mathbf{P}_x(\eta | \zeta_0, \dots, \zeta_n)$$

$$= \lim_{n\to\infty} \mathbf{P}_x(T^n\eta|\zeta_0,\ldots,\zeta_n) = \lim_{n\to\infty} \varphi(\zeta_n)$$

is almost certainly valid with respect to $\mathbf{P}_x$. Moreover,

$$\eta = \varphi(\zeta_n) \tag{39}$$

is also almost certainly valid with respect to $\mathbf{P}_x$. Indeed, because the variable η is measurable with respect to $\zeta_{2n+1}, \zeta_{2n+2}, \ldots$ for any natural n, the variables $\xi_1, \xi_2, \ldots$ are independent and identically distributed, so that:

$$\begin{aligned}\mathbf{P}_x|\eta - \varphi(\zeta_n)| &= \mathbf{P}_x|\eta - \mathbf{P}_x(\eta|\xi_1,\ldots,\xi_n)| \\ &= \mathbf{P}_x|\eta - \mathbf{P}_x(\eta|\xi_{n+1},\ldots,\xi_{2n})|\end{aligned}$$

and, consequently, the sequence $\mathbf{P}_x(\eta|\xi_{n+1},\ldots,\xi_{2n})$, $n = 1,2,\ldots$, converges to η with probability $\mathbf{P}_x$. Since $(\xi_1,\ldots\xi_n)$ and $(\xi_{n+1},\ldots,\xi_{2n})$ are independent;

$$\begin{aligned}\mathbf{P}_x\eta^2 &= \lim_{n\to\infty} \mathbf{P}_x[\mathbf{P}_x(\eta|\xi_1,\ldots,\xi_n)\mathbf{P}_x(\eta|\xi_{n+1},\ldots,\xi_{2n})] \\ &= \lim_{n\to\infty} \{\mathbf{P}_x[\mathbf{P}_x(\eta|\xi_1,\ldots,\xi_n)]\mathbf{P}_x[\mathbf{P}_x(\eta|\xi_{n+1},\ldots,\xi_{2n})]\} \\ &= (\mathbf{P}_x\eta)^2.\end{aligned}$$

Because $\eta = \mathbf{P}_x\eta$ is almost certainly true with respect to $\mathbf{P}_x$ then

$$\lim_{n\to\infty} (\varphi(x + \xi_1 + \ldots + \xi_n (\mathrm{mod}\, 1)) = \varphi(x)$$

is also almost certainly true with respect to $\mathbf{P}_x$.

Therefore if $\xi_1, \xi_2, \ldots$ are independent and identically distributed, then;

$$\eta = \lim_{n\to\infty} \varphi(\zeta_{n+m}) = \lim_{n\to\infty} \varphi(\zeta_m + \xi_{m+1} + \ldots + \xi_{n+m}(\mathrm{mod}\, 1)) = \varphi(\zeta_m)$$

is almost certainly true with respect to $\mathbf{P}_x$.

In particular, $\varphi(x + \xi_1(\mathrm{mod}\, 1)) = \varphi(x)$ is almost certainly true with respect to $\mathbf{P}_x$ and, consequently, if the function $\varphi^\bullet$ is a periodic extension of the function φ from the interval $[0,1)$ to $[0,\infty)$ with a unit period, then:

$$\varphi^\bullet(x + \xi_1) = \varphi^\bullet(x) \tag{40}$$

is almost certainly true with respect to $\mathbf{P}_x$ for all $x \geq 0$ (here, $\mathbf{P}_x = \mathbf{P}_y$ if $x - y = 1$). Together with the function $\varphi^\bullet$, the function:

$$\psi(x) = \int_0^\infty \varphi^\bullet(x + y)e^{-ny}dy, \qquad x \geq 0,$$

also satisfies equation (40) for any natural n. This function is continuous and, like $\varphi^\bullet$, is periodic with a unit period. The continuity of ψ, which follows from the formula;

$$\psi(x) = e^{nx} \int_x^\infty \varphi^\bullet(y)e^{-ny}dy,$$

implies that $\sup_{x\geq 0}|\psi(x+\xi_1)-\psi(x)|=0$ with unit probability. This means that there exists a set $I\subset[0,1)$ with $F(I)=1$ such that every element of I is a period of the function ψ. But, because ψ has already a period equal to unity, and the distribution F is not concentrated on the set $\{0,h,2h,\dots\}$ for any rational h, this can occur only if $\psi=\text{const}$. Thus;

$$\int_0^\infty [\varphi^\bullet(x+y)-\varphi^\bullet(y)]e^{-ny}dy=0$$

for all $x\geq 0$, $n=0,1,2,\dots$. This is possible if and only if $\varphi(x)=\text{const}$ for almost all $x\in[0,1)$ with respect to the Lebesgue measure. Since ζ_n has a uniform distribution with probability $\mathbf{P}^\bullet$, this, together with (39), implies that:

$$\eta=\mathbf{P}^\bullet\eta$$

is almost certainly true with respect to $\mathbf{P}^\bullet$. This implies that the σ-algebra $\mathcal{I}$ of invariant events is trivial with probability $\mathbf{P}^\bullet$. It also implies, together with the possession of a stationary property, that the sequence ξ_n, $n=0,1,2,\dots$, is ergodic with probability $\mathbf{P}^\bullet$. Moreover, $\mathbf{P}^\bullet(\Gamma)=1$ if and only if $\mathbf{P}_x(\Gamma)=1$ for almost all x with respect to the Lebesgue measure.

Hence, if $\mathbf{P}^\bullet|\eta|<\infty$, then:

$$\lim_{n\to\infty}\frac{1}{n}\sum_{k=0}^{n-1}T^k\eta=\mathbf{P}^\bullet\eta$$

is almost certainly true with respect to $\mathbf{P}_x$ for almost all $x\in[0,1)$ with respect to the Lebesgue measure.

26. *Regenerating processes* give an example of a random translation transformation, which is ergodic and preserves the probability. A measurable random process $\mathfrak{z}_t$ $(t\geq 0)$ is called regenerating, with a regenerating time τ, if the processes ζ_t $(t<\tau)$ and $\zeta_{\tau+t}$ $(t\geq 0)$ are independent and the distribution of the latter process coincides with the distribution of the process ζ_t $(t\geq 0)$. It is assumed that $0<\tau<\infty$ with unit probability. Thus, if the time, τ, is measurable with respect to the σ-algebra $\mathcal{M}$, which is generated by the process ζ_t $(t\geq 0)$, then the definition of a regenerating process already implies that the measure $\mathbf{P}$ on $\mathcal{M}$ is invariant with respect to T_τ.

Moreover, if τ is measurable with respect to $\mathcal{M}$, then $\tau_k=T_\tau z_{k-1}$, $k=2,3,\dots$, $\tau_1=\tau$, is a regenerating time for the process ζ_t $(t\geq 0)$. The proof of this is based on the following auxiliary statement.

Let the mapping $(t,\omega)\to\xi_t(\omega)$ be bounded and measurable with respect to $\mathcal{B}_t\times\mathcal{M}$, and let $\eta_t=T_t\xi_t$, then:

$$\mathbf{P}(\zeta_t\in A,t<\tau,\eta_\tau)=\mathbf{P}(\zeta_t\in A,t<\tau,\varphi(\tau)),\tag{41}$$

where $A \in \mathcal{A}$ and $\varphi(u) \in \mathbf{P}\xi_u$.

Indeed, if $\xi_t(\omega) = i_{[0,s)}(t)\xi(\omega)$, where $s > 0$ and the variable ξ is measurable with respect to $\mathcal{M}$, and $\varphi(\tau) = i_{\{\tau<s\}}\mathbf{P}\xi$, then equation (41) becomes a direct consequence of the definition of a regenerating process. In the general case, (41) is obtained from linear combinations of the functions, $\xi_t(\omega)$ $(t \geq 0, \omega \in \Omega)$, and taking the limit.

The event $\{\zeta_t \in A, t < \tau_2\}$ can be written as $\{\zeta_t \in A, t < \tau\} \cup \Gamma_\tau$, where $\Gamma_u = T_u^{-1}\Lambda_u$, $\Lambda_u = \{\zeta_{t-u} \in A, 0 \leq t - u < \tau\}$.

According to formula (41),

$$
\begin{aligned}
\mathbf{P}(\mathfrak{z}_t \in A, t < \tau_2, T_{\tau_2}\xi) &= \mathbf{P}(\mathfrak{z} \in A, t < \tau, T_\tau^2\xi) + \mathbf{P}(\Gamma_\tau, T_\tau^2\xi) \\
&= \mathbf{P}(\mathfrak{z}_t \in A, t < \tau)\mathbf{P}\xi + \mathbf{P}\varphi(\tau),
\end{aligned}
$$

where $\varphi(u) = \mathbf{P}(\Lambda_u, T_\tau\xi) = \mathbf{P}(\Lambda_u)\mathbf{P}\xi$. Consequently,

$$
\mathbf{P}(\mathfrak{z}_t \in A, t < \tau_2, T_\tau^2\xi) = [\mathbf{P}(\mathfrak{z}_t \in A, t < \tau) + \mathbf{P}\chi(\tau)]\mathbf{P}\xi,
$$

where $\chi(u) = \mathbf{P}(\mathfrak{z}_{t-u} \in A, 0 \leq t - u < \tau)$. Because the expression in brackets does not change when ξ changes, the equation obtained demonstrates that the event $\{\mathfrak{z}_t \in A, t < \tau_2\}$ and the variable $T_{\tau_2}\xi$ are independent. The independence of the variable $T_{\tau_2}\xi$ and the event $\{\mathfrak{z}_{t_1} \in A_1, \ldots < \mathfrak{z}_{t_n} \in A_n, t_1 < \ldots < t_n < \tau_2\}$ can be proved in a similar way by using induction with respect to n (this proof will be carried out completely in Section 5 in a situation more general than the one considered here).

The analysis, in the case of an arbitrary time τ_k, differs from the one given only by the notation used, and demonstrates that each τ_k $(k = 1, 2, \ldots)$ possesses the property of regeneration. It shows also, in particular, that the variables τ_1, $\tau_2 - \tau_1$, … , $\tau_n - \tau_{n-1}$, … are independent and identically distributed. Hence τ_k tends to ∞ with a unit probability. Thus, if a measurable (with respect to $\mathcal{M}$) variable, η, is bounded and invariant with respect to T_τ, then:

$$
\begin{aligned}
\eta &= \lim_{n\to\infty} \mathbf{P}(\eta|\mathfrak{z}_u, u < \tau_n) = \lim_{n\to\infty} \mathbf{P}(T_{\tau_n}\eta|\mathfrak{z}_u, u < \tau_n) \\
&= \lim_{n\to\infty} \mathbf{P}(T_{\tau_n}\eta) = \mathbf{P}\eta
\end{aligned}
$$

and, consequently, the transformation T_τ is ergodic.

27. *A branching process with immigration* is an example of an ergodic process for which a translation operator that preserves the probability is not required to exist.

A formal definition of a branching process, in which transformations of particles depend on time, is beyond the scope of this book. However, an informal description is very simple and clear. Every branching process is connected with some population, which evolves in a random way. Each of the particles at a certain time, regardless of their origin, transform into some (possibly empty) collection of new particles. The number of descendants a particle has can depend only on the time at which the transformation occurred.

Moreover, there is an inflow of particles (immigration) into the population taking place at independent, identically distributed, times with identically distributed amounts. Each of the new particles becomes involved in the branching process, in which it participates equally with the particles already existing in the population. The state of a branching process with immigration is a complete collection of the ages of all existing particles as well as the time passed from the last immigration. Trajectories of a branching process are right continuous and the set of trajectories is invariant with respect to translations. Let $p_k(u)$ be the conditional probability that the number of descendants of a particle is equal to k with the condition that the transformation occurred at the time, u ($k = 0, 1, 2, \ldots$, $u \geq 0$), and let $q_k(u)$ ($k = 1, 2, \ldots$, $u \geq 0$) be the conditional distribution of the number of particles which have immigrated, if the time elapsed from the last immigration is equal to u. Suppose also that the distribution of the lifetimes of single particles, $L(du)$, together with the distribution of time between immigrations, $N(du)$, do not have an atom at the origin i.e. $L\{0\} = N\{0\} = 0$. Then, if

$$\int_0^\infty L(du) \sum_{k=0}^\infty k p_k(u) < 1, \qquad \int_0^\infty N(du) \sum_{k=1}^\infty k q_1(u) < \infty,$$

$$\int_0^\infty N(du) u < \infty, \qquad \int_0^\infty L(du) u < \infty,$$

then the branching process with immigration is ergodic.

The proof is based on the existence of the limits;

$$\lim_{t\to\infty} \frac{1}{t} \int_0^t \mathbf{P}_0 T_u \xi du = \mathbf{P}^\bullet \xi, \tag{42}$$

where $\mathbf{P}_0$ is the conditional probability, with respect to the σ-algebra $\mathcal{M}$, generated by the process (under the condition that, initially, the population was empty), and the random variable ξ is bounded and measurable with respect to $\mathcal{M}$.

The limit probability $\mathbf{P}^\bullet$ can be written as:

$$\begin{aligned} \mathbf{P}^\bullet \xi &= \frac{1}{\int_0^\infty N(du) u} \left\{ \int_0^\infty N(du) \int_0^u \mathbf{P}_0 T_t \xi dt \right. \\ &\quad \left. + \int_0^\infty N(du) \int_0^\infty dt [\mathbf{P}_0 T_{t+u} \xi - P_0 T_t \xi] \right\}. \end{aligned} \tag{43}$$

Indeed, the function $\varphi(t) = \mathbf{P}_0 T_t \xi$ ($t \geq 0$) satisfies the equation;

$$\varphi(t) = f(t) + \int_0^t N(du) \varphi(t-u), \tag{44}$$

where;

$$f(t) = \mathbf{P}_0(T_t\xi, t < \tau) + \int_0^t N(du) \sum_{k=1}^{\infty} q_k(u)[P_k T_{t-u}\xi - \mathbf{P}_0 T_{t-u}\xi]. \tag{45}$$

Here, τ is the first immigration time and $\mathbf{P}_k$ is the conditional probability on $\mathcal{M}$, with the condition that the initial set consisted of k new particles.

To prove (44), the formula can be written for the average probability in terms of the first immigration time, τ:

$$\begin{aligned} \varphi(t) &= \mathbf{P}_0(T_t\xi, t < \tau) + \int_0^t N(du) \sum_{k=1}^{\infty} q_k(u)\mathbf{P}_k T_{t-u}\xi \\ &= f(t) + \int_0^t N(du) \sum_{k=1}^{\infty} q_k(u)\mathbf{P}_0 T_{t-u}\xi \\ &= f(t) + \int_0^t N(du)\varphi(t-u). \end{aligned}$$

To prove (42), it is sufficient to show that the function f is integrable. Let $|\xi| \le c$; then:

$$|\mathbf{P}_0(T_t\xi, t < \tau)| \le cN(t, \infty)$$

and:

$$|\mathbf{P}_k T_t\xi - \mathbf{P}_0 T_t\xi| \le 2c\mathbf{P}_k(\mathfrak{t} > t), \tag{46}$$

where $\mathfrak{t}$ is the degeneration time for the initial set.

To prove the last inequality, besides the process studied, two additional branching processes with immigration need to be considered, which have probability laws of evolution the same as the process under consideration. Initially, the number of particles of the first process is equal to k with unit probability and the set of the particles of the second process is empty. Moreover, the population, which corresponds to the first additional process, contains the population of the second process as a part with unit probability. Suppose also that the processes have the same trajectory space, on which the translation operators T_t $(t \ge 0)$ are defined.

The random variables ξ and $\mathfrak{t}$ can be considered as functionals on trajectories of the main branching process with immigration. The same functionals, but calculated on trajectories of the first and the second additional processes, will be denoted by ξ_k, $\mathfrak{t}_k$, and ξ_0, $\mathfrak{t}_0$ respectively (the variable $\mathfrak{t}_0$ is almost certainly equal to zero).

With this notation;

$$\mathbf{P}_k(\mathfrak{t} > t) = \mathbf{P}(\mathfrak{t}_k > t), \quad \mathbf{P}_k T_k\xi = \mathbf{P}T_t\xi_k, \quad \mathbf{P}_0 T_t\xi = \mathbf{P}T_t\xi_0. \tag{47}$$

The assumptions made imply that the trajectories of the first and the second additional processes coincide after a time $\mathfrak{t}_k$. Consequently;

$$T_t\xi_k = T_t\xi_0 \qquad \text{for} \qquad t \ge t_k.$$

Therefore;

$$\mathbf{P}T_k\xi_k - \mathbf{P}T_t\xi_0 = \mathbf{P}[(T_t(\xi_k - \xi_0), t < \mathbf{t}_k].$$

This, together with (47), gives (46).

The distribution of the random variable $\mathbf{t}_k$ is equal to the distribution of the maximum of k independent random variables, each of which has the same distribution as $\mathbf{t}_1$. So;

$$\mathbf{P}_k(\mathbf{t} > t) \geq k\mathbf{P}_1(\mathbf{t} > t)$$

and the absolute value of the second term in (45) does not exceed:

$$2c\int_0^t N(du)\sum_{k=1}^{\infty} q_k(u)kg(t-u),$$

where $g(t) = \mathbf{P}_1(\mathbf{t} > t)$. To prove the integrability of the function f it will suffice to demonstrate that the mean $\mathbf{P}_1\mathbf{t}$ is finite (which is the same as the integrability of the function g). By using the formula for the average probability;

$$\begin{aligned}\mathbf{P}_1(\mathbf{t} \leq t) &= \int_0^t L(du)\sum_{k=0}^{\infty} p_k(u)\mathbf{P}_k(\mathbf{t} \leq t-u) \\ &= \int_0^t L(du)\sum_{k=0}^{\infty} p_k(u)[\mathbf{P}_1(\mathbf{t} \leq t)]^k,\end{aligned}$$

which can be written as;

$$1 - g(t) = \int_0^t L(du)\sum_{k=0}^{\infty} p_k(u)[1 - g(t-u)]^k,$$

$$g(t) = L(t,\infty) + \int_0^t L(du)\sum_{k=0}^{\infty} p_k(u)\{1 - [1 - g(t-u)]^k\}$$

and, consequently;

$$g(t) \leq L(t,\infty) + \int_0^t L(du)\sum_{k=1}^{\infty} kp_k(u)g(t-u).$$

Because $\int_0^{\infty} L(du)\sum_{k=1}^{\infty} kp_k(u) < 1$, the former inequality implies that:

$$\int_0^{\infty} g(t)dt \leq \frac{\int_0^{\infty} ul(du)}{1 - \int_0^{\infty} l(du)\sum_{k=1}^{\infty} kp_k(u)}.$$

This proves that limit (42) exists and is equal to:

$$\frac{1}{\int_0^{\infty} uN(du)}\int_0^{\infty} f(t)dt,$$

which, according to the identity;

$$f(t) = \mathbf{P}_0T_t\xi - \int_0^t N(du)\mathbf{P}_0T_{t-u}\xi$$

is equal to the right hand side of (43).

The constructed probability $\mathbf{P}^\bullet$ is invariant with respect to the translation operators, T_t ($t \geq 0$), and coincides with $\mathbf{P}_0$ on the σ-algebra $\mathcal{I}$ of the subsets of $\mathcal{M}$, which are invariant with respect to the translation operators T_t ($t \geq 0$).

To prove that the branching process with immigration is ergodic with respect to the probability $\mathbf{P}_0$, it remains to show that the σ-algebra $\mathcal{I}$ is trivial with respect to $\mathbf{P}_0$.

Let τ_k denote the time of the kth immigration and ζ_k be the time at which the descendants of all the particles, which immigrated at time τ_k, degenerate. Then $\sigma_k = \max(\tau_1 + \zeta_1, \tau_2 + \zeta_2, \ldots, \tau_k + \zeta_k)$ is equal to the time at which the branching process with immigration will not contain any descendants of the particles that immigrated at the times $\tau_1, \tau_2, \ldots \tau_k$. Thus, if an event Λ is determined by the development of the branching process with immigration after the time σ_k, then Λ does not depend on the σ-algebra $\mathcal{M}_{\tau_k}$ generated by the development of the process up to the time τ_k.

Let a measurable (with respect to $\mathcal{I}$) variable, η, be bounded. By Levy's theorem;

$$\begin{aligned}
\mathbf{P}_0\eta^2 &= \mathbf{P}_0[\eta \lim_{k\to\infty} \mathbf{P}_0(\eta|\mathcal{M}_{\tau_k})] \\
&= \lim_{k\to\infty} \mathbf{P}_0[\eta \mathbf{P}_0(\eta|\mathcal{M}_{\tau_k})] = \lim_{k\to\infty} \mathbf{P}_0[(T_{\sigma_k}\eta)\mathbf{P}_0(\eta|\mathcal{M}_{\tau_k})] \\
&= \lim_{k\to\infty} \mathbf{P}_0(T_{\sigma_k}\eta)\mathbf{P}_0[\mathbf{P}_0(\eta|\mathcal{M}_{\tau_k})] = (\mathbf{P}_0\eta)^2,
\end{aligned}$$

which is possible only if $\eta = \mathbf{P}_0\eta$ is almost certainly valid with respect to $\mathbf{P}_0$. Hence;

$$\lim_{t\to\infty} \frac{1}{t}\int_0^t T_u\xi du = \mathbf{P}^\bullet\xi \tag{48}$$

is almost certainly valid with respect to $\mathbf{P}_0$. If the initial set of particles is not empty, then the time, t, at which the descendants of the initial particles disappear, is almost certainly finite, and, consequently;

$$\begin{aligned}
\lim_{t\to\infty} \frac{1}{t}\int_0^t T_u\xi du &= \lim_{t\to\infty} \frac{1}{t}\int_{\mathrm{t}}^{t+\mathrm{t}} T_u\xi du \\
&= \lim_{t\to\infty} \frac{1}{t}\int_0^t T_u T_{\mathrm{t}}\xi du
\end{aligned}$$

is almost certainly true. The random variable under the limit sign in the last part of the equation does not depend on the initial set. Hence, the initial set can be regarded as being empty with unit probability, and so (48) is almost certainly valid with respect to $\mathbf{P}$.

Chapter 2

Markovian processes

Saying that a random process possesses the Markovian property implies that the future of the process is conditionally independent from the past if the present state of the process is given. If a process possesses the Markovian property at a given time (the stopping time), $\mathfrak{m}$, then the random process is called Markovian at that time. Under very general assumptions, it is possible, from the time $\mathfrak{m}$, to construct a monotone unbounded sequence of times $\mathfrak{m}_k$, $k = 1, 2, \ldots$, each of which has the Markovian property. Values of the process at these times form, what is called a imbedded Markovian chain, which define the ergodic properties of the process. The existence of an invariant initial distribution of the imbedded Markovian chain implies that limits of the time means exist, and recurrence of the imbedded Markovian chain insures that these limits are degenerate.

2.1 Recurrent Markovian chains

28. A sequence of random variables, $\mathfrak{z}_n$ $(n = 0, 1, 2, \ldots)$, taking values in a phase space $(E, \mathcal{A})$, is called a *Markovian chain* (more precisely, a homogeneous Markovian chain) if the conditional probabilities $\mathbf{P}_x(\cdot) = \mathbf{P}(\cdot|\mathfrak{z}_0 = x)$ are regular on the σ-algebra $\mathcal{N}$, which is generated by the sequence $\mathfrak{z}_n$, $n = 0, 1, 2, \ldots$, and if:

$$\mathbf{P}_x(\mathfrak{z}_{n+1} \in A|\mathfrak{z}_0, \ldots, \mathfrak{z}_n) = \mathbf{P}_{\mathfrak{z}_n}(\mathfrak{z}_1 \in A) \tag{49}$$

is almost certainly true, with respect to $\mathbf{P}_x$, for all $x \in E$ $(n = 0, 1, 2, \ldots, A \in \mathcal{A})$.

If the set of trajectories of the Markovian chain ,$\mathfrak{z}_n$ $(n = 0, 1, \ldots)$, is closed with respect to translations (and only those chains are interesting from the point of view of ergodic theory), then there is an operator T, which is defined on measurable (with respect to $\mathcal{N}$) variables, which is a translation of the sequence $\mathfrak{z}_n$ $(n = 0, 1, \ldots)$ by one value of n and so definition (49) is equivalent to the following:

$$\mathbf{P}_x(\xi T^n \eta) = \mathbf{P}_x(\xi \mathbf{P}_{\mathfrak{z}_n} \eta), \tag{50}$$

where $x \in E$ $(n = 1, 2, \ldots)$, the variable η is bounded and measurable with respect to $\mathcal{N}$ and the bounded variable ξ is measurable with respect to the σ-algebra $\mathcal{N}_n$, which is the minimal σ-algebra with respect to which $\mathfrak{z}_0, \mathfrak{z}_1, \ldots, \mathfrak{z}_n$ are measurable.

Formula (50) follows from (49) if $\eta = \varphi_0(\mathfrak{z}_0)\varphi_1(\mathfrak{z}_1)\ldots\varphi_k(\mathfrak{z}_k)$, where the functions $\varphi_0, \ldots, \varphi_k$ are bounded and measurable with respect to A. Linear combinations of such functions approximate, after an at most a countable number of translations to the limit, in the sense of a pointwise convergence, all bounded measurable (with respect to $\mathcal{A}^{k+1}$) functions, φ, on E^{k+1}. Thus (50) holds for $\eta = \varphi(\mathfrak{z}_0, \ldots, \mathfrak{z}_k)$ and, consequently, for all measurable (with respect to $\mathcal{N}$) bounded variables.

29. The Markovian property given in (49) and (50) is also preserved at some random times: the stopping times. A nonnegative integer variable, τ, is a stopping time for a sequence $\mathfrak{z}_n$, $n = 0, 1, 2, \ldots$, if $\{\tau = n\} \in \mathcal{N}_n$ for all $n = 0, 1, 2, \ldots$. Let τ be a stopping time for a Markovian chain, $\mathfrak{z}_n$ $(n = 0, 1, 2, \ldots)$, and $\mathbf{P}_x(\tau < \infty) \equiv 1$. Then;

$$\mathbf{P}_x(\mathfrak{z}_{\tau+1} \in A | \mathcal{N}_\tau) = \mathbf{P}_{\mathfrak{z}_\tau}(\mathfrak{z}_1 \in A)$$

is almost certainly valid, with respect to $\mathbf{P}_x$, for all $x \in E$ $(A \in \mathcal{A})$, which can also be written as;

$$\mathbf{P}_x(\xi T_\tau \eta) = \mathbf{P}_x(\xi \mathbf{P}_{\mathfrak{z}_\tau} \eta), \tag{51}$$

where $x \in E$ and the variable η is bounded and measurable with respect to $\mathcal{N}$. $T_\tau \eta(\omega) = T^k \eta(\omega)$ if $\tau(\omega) = k$. The bounded variable ξ is measurable with respect to the σ-algebra $\mathcal{N}_\tau$ consisting of measurable (with respect to $\mathcal{N}$) events, Γ, such that $\Gamma \cap \{\tau = n\} \in \mathcal{N}_n$ for all $n = 0, 1, 2, \ldots$. Formula (51) follows from (50) and the fact that the variable ξ is measurable with respect to $\mathcal{N}_k$ on the set $\{\tau = k\}$. Indeed, the left-hand side of (51) equals $\sum_{k=0}^{\infty} \mathbf{P}_x(\tau = k, \xi T^k \eta)$. The kth term here equals $\mathbf{P}_x(\tau = k, \xi \mathbf{P}_{\mathfrak{z}_k} \eta)$ which, being summed for all $k = 0, 1, 2, \ldots$, gives the right-hand side of (51).

30. The collection of probabilities P_x $(x \in E)$ on $\mathcal{N}$ is uniquely defined by the transition probability $\mathbf{P}_x(B) = \mathbf{P}_x(\mathfrak{z}_1 \in B)$ by one step for the corresponding chain. Indeed, by formula (49), for $n = 1$;

$$\begin{aligned}\mathbf{P}_x(\mathfrak{z}_1 \in B_1, \mathfrak{z}_2 \in B_2) &= \mathbf{P}_x(\mathfrak{z}_1 \in B_1, \mathbf{P}_{\mathfrak{z}_1}(\mathfrak{z}_1 \in B_2)) \\ &= \int_{B_1} \mathbf{P}_x(dy)\mathbf{P}_y(B_2),\end{aligned}$$

which, together with (49) for $n = 2$, yields the following:

$$\begin{aligned}\mathbf{P}_x(\mathfrak{z}_1 \in B_1, \mathfrak{z}_2 \in B_2, \mathfrak{z}_3 \in B_3) &= \mathbf{P}_x(\mathfrak{z}_1 \in B_1, \mathfrak{z}_2 \in B_2, \mathbf{P}_{\mathfrak{z}_2}(\mathfrak{z}_1 \in B_3)) \\ &= \mathbf{P}_x(\mathfrak{z}_1 \in B_1, \mathbf{P}_{\mathfrak{z}_1}(\mathfrak{z}_1 \in B_2, \mathbf{P}_{\mathfrak{z}_1}(\mathfrak{z}_1 \in B_3)))\end{aligned}$$

$$= \int_{B_1} \mathbf{P}_x(dy_1) \int_{B_2} \mathbf{P}_{y_1}(dy_2)\mathbf{P}_{y_2}(B_3),$$
$$\mathbf{P}_x(\mathfrak{z}_1 \in B_1, \ldots, \mathfrak{z}_n \in B_n)$$
$$= \int_{B_1} \mathbf{P}_x(dy_1) \ldots \int_{B_k} \mathbf{P}_{y_k}(dy_k) \ldots \int_{B_{n-1}} \mathbf{P}_{y_{n-1}}(B_n).$$

This implies, in particular, that the k-step transition probabilities, $P_x^k(dy)$, for the chain satisfy the relation;

$$\int \mathbf{P}_x^k(dy)\mathbf{P}_y^l(B) = \mathbf{P}_x^{k+l}(B),$$

also known as *Kolmogorov-Chapman's equation.*

31. In the framework of Markovian chains, it is reasonable to narrow the definition of the ergodicity of random processes given in Section 2. Namely, a *Markovian chain* $\mathfrak{z}_n$, $n = 0, 1, 2, \ldots$, *is called ergodic* if for all $x \in E$ and all measurable (with respect to $\mathcal{N}$) bounded variables ξ, there almost certainly exists, with respect to $\mathbf{P}_x$, the nonrandom limit;

$$\lim_{n\to\infty} \frac{1}{n} \sum_{k=0}^{n-1} T^k\xi, \tag{52}$$

which does not depend on x. If a Markovian chain $\mathfrak{z}_n$, $n = 0, 1, 2, \ldots$, is ergodic, then limit (52) becomes $\mathbf{P}_\pi\xi = \int \pi(dx)\mathbf{P}_x\xi$, where the probability distribution π on $(E, \mathcal{A})$ is invariant, i.e. $\mathbf{P}_\pi(\mathfrak{z}_1 \in A) = \pi(A)$, or, in terms of the transition probability $P_x(dy)$;

$$\int \pi(dx)\mathbf{P}_x(A) = \pi(A), \qquad A \in \mathcal{A}. \tag{53}$$

Indeed, for an ergodic chain, limit (52) can be written as:

$$\lim_{n\to\infty} \frac{1}{n} \sum_{k=0}^{n-1} \mathbf{P}_x T^k\xi. \tag{54}$$

Because the chain is Markovian, $\mathbf{P}_x T^k\xi = \mathbf{P}_x\varphi(\mathfrak{z}_k)$, where $\varphi(x) = \mathbf{P}_x\xi$. However, for $\xi = \varphi(\mathfrak{z}_0)$, limits (52) and (54) are equal to $\int \pi(dx)\varphi(x)$, where;

$$\pi(A) = \lim_{n\to\infty} \frac{1}{n} \sum_{k=0}^{n-1} \mathbf{P}_x(\mathfrak{z}_k \in A), \qquad A \in \mathcal{A}$$

and, consequently, limit (52) equals $\mathbf{P}_\pi\xi$. Moreover, the limits given by (52) coincide for $\xi = \varphi(\mathfrak{z}_0)$ and $\xi = \varphi(\mathfrak{z}_1)$. Hence $\mathbf{P}_\pi\varphi(\mathfrak{z}_1) = \mathbf{P}_\pi\varphi(\mathfrak{z}_0) = \int \pi(dx)\varphi(x)$, which is equivalent to (53).

Therefore, for an ergodic Markovian chain, $\mathfrak{z}_n$ $(n = 0, 1, 2, \ldots)$, there exists an invariant probability distribution π on $\mathcal{A}$ and:

$$\lim_{n\to\infty} \frac{1}{n} \sum_{k=0}^{n-1} i_A(\mathfrak{z}_k) = \pi(A)$$

is almost certainly true, with respect to $\mathbf{P}_x$, for all $x \in E$ $(A \in \mathcal{A})$. Consequently, if $\pi(A) > 0$, then the probability, $\mathbf{P}_x$, of the event, $\{\mathfrak{z}_n \in A$, for an infinite number of $n\}$, is equal to unity for all $x \in E$. This property is called the *(Harris) π-recurrence* of the chain $\mathfrak{z}_n$, $n = 0, 1, \ldots$.

32. Conversely, the existence of an invariant probability distribution, π, and the recurrence (with respect to μ) of some nontrivial σ-finite measure μ, which is absolutely continuous with respect to π, are sufficient conditions for a Markovian chain, $\mathfrak{z}_n$ $(n = 0, 1, 2, \ldots)$, to be ergodic. Indeed, the translation operator T preserves the measure $\mathbf{P}_\pi$ on $\mathcal{N}$; $\mathbf{P}T\xi = \mathbf{P}_\pi \mathbf{P}_{\mathfrak{z}_1}\xi = \mathbf{P}_\pi\xi$. Thus, if ξ is bounded and measurable with respect to $\mathcal{N}$, then the sequence $\xi_k = T^k\xi$ is stationary with probability $\mathbf{P}_\pi$ and uniformly bounded. By Corollary 7, the limit $\lim_{n\to\infty} 1/2 \sum_{k=0}^{n-1} T^k\xi = \zeta$ almost certainly exists with respect to $\mathbf{P}_\pi$, with $\mathbf{P}_\pi\zeta = \mathbf{P}_\pi\xi$. The variable ζ can be chosen so that it is measurable with respect to the σ-algebra $\mathcal{I}$ of the events from $\mathcal{N}$ that are invariant with respect to T.

If a measurable (with respect to $\mathcal{I}$) variable η is bounded, then by Levy's Theorem;

$$\begin{aligned} \eta &= \mathbf{P}_x(\eta|\mathcal{N}) = \lim_{n\to\infty} P_x(\eta|\mathcal{N}_n) \\ &= \lim_{n\to\infty} \mathbf{P}_x(T^{n+1}\eta|\mathcal{N}_n) \\ &= \lim_{n\to\infty} \varphi(\mathfrak{z}_n) \end{aligned}$$

is almost certainly true with respect to $\mathbf{P}_x$, where $\varphi(y) = \mathbf{P}_y\eta$.

However, for a recurrent (with respect to μ) chain, the function φ must be constant because, if the measure, μ, of the set $C = \{x \in E : a < \varphi(x) < b\}$ is positive, then the time τ_C, at which the chain $\mathfrak{z}_n$ $(n = 0, 1, 2, \ldots)$ reaches the set C for the first time, is almost certainly finite with respect to $\mathbf{P}_x$. Because η is invariant with respect to T, together with $\tau = \tau_C$, then from (51);

$$\varphi(x) = \mathbf{P}_x\eta = \mathbf{P}_x T_{\tau_C}\eta = \mathbf{P}_x\varphi(\mathfrak{z}_{\tau_C}) \in (a, b)$$

for all $x \in E$. Thus $\eta = \mathbf{P}_\pi\eta$ is almost certainly true with respect to $\mathbf{P}_x$, which implies that the σ-algebra, $\mathcal{I}$, is trivial with respect to every probability $\mathbf{P}_x$, where $x \in E$.

Consequently;

$$\lim_{n\to\infty} \frac{1}{n} \sum_{k=0}^{n-1} T^k\xi = \mathbf{P}_\pi\xi \tag{55}$$

is almost certainly true with respect to P_π. Let the event Γ exist such that the left hand side limit exists and is equal to the right hand side. If $C = \{x \in E : P_x(\Gamma) = 1\}$

then $\pi(C) = 1$ and consequently, $\mu(C) > 0$. Hence, $\mathbf{P}_x\{\tau_C < \infty\} = 1$. Because $\Gamma \in \mathcal{I}$;

$$\mathbf{P}_x(\Gamma) = \mathbf{P}_x(T_{\tau_C}^{-1}\Gamma) = \mathbf{P}_x\mathbf{P}_{\mathfrak{z}_{\tau_C}}(\Gamma) = 1$$

which completes the proof.

33. A Markovian chain is called *recurrent* if it is recurrent, with respect to μ, for a suitable choice of a σ-finite nontrivial measure μ.

A measure π on $\mathcal{A}$ is called invariant for a Markovian chain $\mathfrak{z}_n$ $(n = 0, 1, \ldots)$ if:

$$\mathbf{P}_\pi(\mathfrak{z}_1 \in B) = \int \pi(dx)\mathbf{P}_x(\mathfrak{z}_1 \in B) = \pi(B)$$

for $\pi(B) < \infty$ $(B \in \mathcal{A})$.

If only the σ-algebra $\mathcal{A}$ is countably generated, a recurrent chain has an invariant σ-finite measure π which is also recurrent.

Theorem. *Let a homogeneous chain $\mathfrak{z}_n$ $(n = 0, 1, 2, \ldots)$ in a phase space $(E, \mathcal{A})$, with the countably generated σ-algebra $\mathcal{A}$, be recurrent. Then;*

the chain $\mathfrak{z}_n$, $n = 0, 1, 2, \ldots$, has a σ-finite invariant measure π and is recurrent with respect to π;

the measure μ, which makes the chain recurrent,is absolutely continuous with respect to π and;

the measure μ is unique in the class of invariant measures such that any measure that satisfies the above conditions is a constant multiple of μ.

Proof. Because the σ-algebra, $\mathcal{A}$, is countably generated, the density of the absolutely continuous component (with respect to the measure μ, which makes the chain recurrent) of the k-step transition probability, $P_x^k(dy)$, of the Markovian chain $\mathfrak{z}_n$ $(n = 0, 1, \ldots)$ is measurable with respect to the pair of arguments, i.e.

$$\mathbf{P}_x^k(B) = \int_B p_k(x, y)\mu(dy) + \lambda_x^k(B),$$

where the measure λ_x^k is singular with respect to μ and the function p_k is nonnegative and measurable with respect to $\mathcal{A} \times \mathcal{A}$. Here, without loss of generality, one can assume that the measure μ is a probability measure. Then, for a given positive number, ϵ, there exists a natural number, m, and a measurable (with respect to $\mathcal{A}$) set, C, such that:

$$\mu(CB_x^m) \geq \epsilon \tag{56}$$

for all $x \in C$, where $B_x^m = \{y : \sum_{k=1}^m p_k(x, y) > 1/r\}$ (since p_k is measurable with respect to $\mathcal{A} \times \mathcal{A}$. It follows that B_x^r is measurable with respect to $\mathcal{A}$). Indeed, let the set D_x be such that $\mu(D_x) = 0$ and $\lambda_x^k(\rceil D_x) = 0$. Then, because $p_k(x,y) = 0$ for $y \notin \cup_{r=1}^{\infty} B_x^r$, $\mathbf{P}_x^k(\rceil \cup_{r=1}^{\infty} (B_x^r \cup D_x)) = 0$ and, consequently, $\mathbf{P}_x(\mathfrak{z}_n \notin \cup_{r=1}^{\infty} B_x^r \cup D_x$ for some $n = 1,2,\ldots) = 0$. This implies, because the chain is recurrent with respect to μ, that $\mu(\rceil \cup_{r=1}^{\infty} B_x^r) = \mu((\rceil \cup_{r=1}^{\infty} B_x^r)\rceil D_x) = 0$. Consequently, $\lim_{r\to\infty} \mu(B_x^r) = \mu(E) = 1$. Moreover, $\mu(B_x^r)$, as a function of x, is measurable with respect to $\mathcal{A}$, and this allows the definition of a finite measurable, with respect to $\mathcal{A}$, integer function $n(x) = \min\{k \geq 1 : \mu \times (B_x^k) \geq 1 - \epsilon/2\}$. If $m = \min\{k : \mu\{x : n(x) \leq k\} \geq 3/2\epsilon\}$ and $C = \{x : n(x) \leq m\}$, then $\mu(CB_x^m) = \mu(B_x^m) - \mu(B_x^m \rceil C) \geq 1 - \epsilon/2 - \mu(\rceil C) = \mu(C) - \epsilon/2 \geq \epsilon$.

Suppose that $\tau_1, \tau_2, \ldots$ are consecutive times at which the chain $\mathfrak{z}_n$ $(n = 1,2,\ldots)$ intersects the set C, i.e. $\tau_1 = \tau_C = \min\{k \geq 1 : \mathfrak{z}_k \in C\}$ $(\tau_{k+1} = \tau_1 + T_{\tau_1}\tau_k$, $k = 1,2,\ldots)$. Since $\mu(C) > 0$, $\tau_k < \infty$ is almost certainly true with respect to $\mathbf{P}_x$. Moreover, the variable τ_k is a stopping time of the chain $\mathfrak{z}_n$ $(n = 0,1,2\ldots)$ for any $k = 1,2,\ldots$. So, by (51) and the fact that $\mathfrak{z}_{\tau_k} = T_{\tau_1}^k \mathfrak{z}_0$, the sequence $\mathfrak{h}_n$ $(n = 0,1,2,\ldots)$, where $\mathfrak{h}_n = \mathfrak{z}_\tau$ $(n \geq 1)$, is a homogeneous Markovian chain in the phase space $(C, \mathcal{A}_C)$, with P_x $(x \in C)$ as conditional probabilities (under the condition that $\mathfrak{h}_0 = x$). The k-step transition probabilities $Q_x^k(dy)$ for this chain satisfy the inequality;

$$\sum_{k=1}^{m} Q_x^k(A) \geq \sum_{k=1}^{m} \mathbf{P}_x^k(A), \quad x \in C, \quad A \in \mathcal{A}_C,$$

which can be proved by noticing that the left hand side is the mean value of the number of times the chain $\mathfrak{z}_k$ $(k = 1,2,\ldots)$ visits the subset A of the set C in the first m visits of the set C, and the right hand side is the mean value of the number of visits of the set A in m steps. Consequently, if the following definition is used:

$$R_x(A) = \frac{1}{m}\sum_{k=1}^{m} Q_x^k(A), \qquad r(x,y) = \frac{1}{m}\sum_{k=1}^{m} p_x(x,y),$$

and (56) is used with $\epsilon = 3/4$, then, for all $x \in C$ and $A \in \mathcal{A}$;

$$R_x(A) \geq \int_A r(x,y)\mu(dy), \qquad \mu\left\{y \in C : r(x,y) \geq \frac{1}{m^2}\right\} \geq \frac{3}{4},$$

which proves the existence of an invariant distribution ρ for the chain $\mathfrak{h}_n$ $(n = 0,1,\ldots)$. Indeed, let:

$$R_x^1(A) = R_x(A),$$
$$R_x^{k+1}(A) = \int_C R_x(dy)R_y^k(A), \qquad k = 1,2,\ldots, \ A \in \mathcal{A}_C,$$
$$D_n = D_n(A) = \{(u,v) \in C \times C : R_u^n(A) > R_v^n(A)\},$$
$$\delta_n = \delta_n(A) = \sup_{x,y\in C}[R_x^n(A) - R_y^n(A)],$$

$$t = \sup_n \sup_{x,y \in C} \iint_{D_n} R_x(du) R_y(dv).$$

Therefore;

$$\begin{aligned} R_x^{n+1}(A) - R_y^{n+1}(A) &= \iint_{C^2} R_x(du) R_y(dv) |R_u^n(A) - R_v^n(A)| \\ &\leq \iint_{D_n} R_x(du) R_y(dv) |R_u^n(A) - R_v^n(A)| \end{aligned}$$

and, consequently, $\delta_{n+1} \leq \delta_n t$, $\delta_{n+1} \leq t^n$. Further;

$$\begin{aligned} \iint_{D_n} R_x(du) R_y(dv) &= 1 - \iint_{C^2 \rceil D_n} R_x(du) R_y(dv) \\ &\leq 1 - \iint_{C^2 \rceil D_n} r(x,y) r(y,v) \mu(du) \mu(dv) \\ &\leq 1 - \iint_{C^2 (B_x^m \times B_y^m) \rceil D_n} r(x,u) r(y,v) \mu(du) \mu(dv) \\ &\quad \text{(because } r(x,u) \geq \frac{1}{m^2} \text{ for } u \in B_x^m \text{)} \\ &\leq 1 - m^{-4} \iint_{C^2 (B_x^m \times B_y^m)} \mu(du) \mu(dv) + m^{-4} \iint_{D_n} \mu(du) \mu(dv) \\ &\leq 1 - \frac{9}{16} m^{-4} + m^{-4} \iint_{D_n} \mu(du) \mu(dv) \leq 1 - (2m)^{-4}. \end{aligned}$$

Here the inequality;

$$\iint_{D_n} \mu(du) \mu(dv) \leq \frac{1}{2},$$

was used, which follows from the fact that the sets D_n and $D_n^- = \{(u,v) : (v,u) \in D_n\}$ do not intersect each other and:

$$\iint_{D_n^-} \mu(dv) \mu(du) = \iint_{D_n} \mu(du) \mu(dv).$$

Hence, $0 \leq t \leq 1$ and:

$$\sup_{x,y \in C \ A \in \mathcal{A}_C} |R_x^n(A) - R_y^n(A)| \leq t^{n-1}. \tag{57}$$

This shows that:

$$|R_x^{n+k}(A) - R_x^n(A)| \leq \int_C R_x^k(dy) |R_y^n(A) - R_x^n(A)| \leq t^{n-1},$$

which, in its turn, (together with (57)) implies that the limit;

$$\rho(A) = \lim_{n \to \infty} R_x^n(A)$$

exists and does not depend on x. The nonnegative function on $\mathcal{A}_C$, ρ, is a measure because it is a limit of the sequence of measures that are convergent everywhere, and:

$$\int_C \rho(dx)R_x(A) = \rho(A), \qquad (58)$$
$$|R_x^n(A) - \rho(A)| \leq t^{n-1}.$$

The last inequality insures that the probability measure on $\mathcal{A}_C$, ρ, satisfying (58) is unique. But, as the identity;

$$\int_C Q_x(dy)R_y(A) = \int_C R_x(dy)Q_y(A)$$

shows, the measure, $\rho^\bullet(A)$, such that $\rho^\bullet(A) = \int_C \rho(dx)Q_x(A)$, together with ρ, satisfies the invariance relation (58). This implies that $\rho^\bullet = \rho$, i.e. ρ is an invariant distribution of the Markovian chain $\mathfrak{y}_n$, $n = 0, 1, 2, \ldots$. If the distribution, ρ, is naturally extended over $\mathcal{A}$, $\rho(A) = \rho(CA)$ $(A \in \mathcal{A})$, then the invariant measure, π, for the chain, $\mathfrak{z}_n$ $(n = 0, 1, 2, \ldots)$, can be defined by the formula:

$$\pi(A) = \mathbf{P}_0 \sum_{k=0}^{\tau_C - 1} i_A(\mathfrak{z}_k), \qquad A \in \mathcal{A}. \qquad (59)$$

Indeed, because $\mathbf{P}_\rho(\mathfrak{z}_{\tau_C} \in A) = \mathbf{P}_\rho(\mathfrak{z}_{\tau_C} \in CA) = \rho(CA) = \mathbf{P}_\rho(\mathfrak{z}_0 \in A)$, then (in the case where $\pi(A) < \infty$);

$$\begin{aligned} \pi(A) &= \mathbf{P}_\rho \sum_{k=1}^{\tau_C} i_A(\mathfrak{z}_k) \\ &= \sum_{k=1}^{\infty} \mathbf{P}_\rho(\tau_C \geq k,\, \mathfrak{z}_k \in A) \\ &= \sum_{k=1}^{\infty} \mathbf{P}_\rho(\tau_C > k-1,\, T^{k-1}\mathfrak{z}_1 \in A) \end{aligned}$$

(because of (50) and since the event $\tau_C > k-1$ is measurable with respect to $\mathcal{N}_{k-1}$)

$$\begin{aligned} &= \sum_{k=1}^{\infty} \mathbf{P}_\rho(\tau_C > k-1,\, \mathbf{P}_{\mathfrak{z}_{k-1}}(\mathfrak{z}_1 \in A)) \\ &= \sum_{k=0}^{\infty} \int \mathbf{P}_\rho(\tau_C > k,\, \mathfrak{z}_k \in dy)\mathbf{P}_y(A) \\ &= \int \pi(dy)\mathbf{P}_y(A). \end{aligned}$$

Moreover, if $\mathbf{P}_\pi \xi < \infty$, then;

$$\mathbf{P}_\pi \xi = \mathbf{P}_\rho \sum_{k=0}^{\tau_C - 1} T^k \xi.$$

Indeed, the right hand side of this equation is equal to $\sum_{k=0}^{\infty} \mathbf{P}_\rho \rho(\tau_C > k, T^k\xi)$, which, according to (50) and the measurability (with respect to $\mathcal{N}_k$) of the event $\{\tau_C > k\}$, coincides with;

$$\sum_{k=0}^{\infty} \mathbf{P}_\rho(\tau_C > k, \mathbf{P}_{\mathfrak{z}_k}\xi) = \mathbf{P}_\rho \sum_{k=0}^{\tau_C - 1} \mathbf{P}_{\mathfrak{z}_k}\xi = \mathbf{P}_\pi \xi.$$

If $\xi = i_{\{\tau_C \le n\}}$ then $T^k \xi = i_{\{\tau_C < k+n\}}$ for $k < \tau_C$ and, consequently, $\mathbf{P}_\pi(\tau_C \le n) = \mathbf{P}_\rho \sum_{k=0}^{\tau_C - 1} i_{\{k > \tau_C - n\}} \le n$ so that the measure $\mathbf{P}_\rho$ is σ-finite on $\mathcal{N}$. Hence, there exists a positive measurable (with respect to $\mathcal{N}$) integrable (with respect to $\mathbf{P}_\pi$) variable, ξ. The function $\varphi(x) = \mathbf{P}_x \xi$, $x \in E$, is positive and integrable with respect to π, i.e. the measure π is σ-finite.

Because the measure, π, coincides with ρ on $\mathcal{A}_C$, it is nontrivial.

To demonstrate that μ is absolutely continuous with respect to π, the following analysis is applied. If $\pi(A) = 0$ then $\mathbf{P}_\pi(\mathfrak{z}_k \in A) = 0$ for all $k = 1, 2, \ldots$. Hence, $\mathbf{P}_\pi \sum_{k=1}^{\infty} i_A(\mathfrak{z}_k) = 0$ and, consequently, the set of those x from E for which $\mathbf{P}_x(\mathfrak{z}_k \in A) = 0$ for some $k = 1, 2, \ldots$ is nonempty. This implies that $\mu(A) = 0$ because μ is recurrent. Because the chain $\mathfrak{y}_n = \mathfrak{z}_{\tau_n}$ $(n = 1, 2, \ldots)$ has an invariant probability distribution, ρ, and it is recurrent with respect to the restriction of μ on $\mathcal{A}_C$, the transformation T_{τ_C} is ergodic, i.e.

$$\lim_{n \to \infty} \frac{1}{n} \sum_{k=0}^{n-1} T_{\tau_C}^k \xi = \mathbf{P}_\rho \xi \tag{60}$$

is almost certainly true with respect to $\mathbf{P}_x$ for all $x \in C$ and integrable (with respect to $\mathbf{P}_\rho$) measurable (with respect to $\mathcal{N}$) variables, ξ.

If $\xi = \sum_{k=0}^{\tau_C - 1} i_A(\mathfrak{z}_k)$ in (60), then;

$$\lim_{n \to \infty} \frac{1}{n} \sum_{k=0}^{\tau_n - 1} i_A(\mathfrak{z}_k) = \pi(A)$$

is almost certainly true with respect to $\mathbf{P}_x$ $(x \in C)$, if $\pi(A) < \infty$. Hence, if $0 < \pi(A) < \infty$, then $\mathbf{P}_x(\tau_A < \infty) = 1$ for all $x \in C$ and, because C is accessible, it is true for all $x \in E$. In the case where $\pi(A) = \infty$; there exists $A^\bullet \subset A$ such that $0 < \pi(A^\bullet) < \infty$, because π is σ-finite.

Finally, consider the invariant σ-finite measure π. The measure, μ, which makes the chain recurrent, is absolutely continuous with respect to π. The uniqueness of π can be obtained in the following way. There exists a measurable (with respect to $\mathcal{A}$) set C such that $0 < \pi(C) < \infty$ and $\mu(C) > 0$. Consequently, $\mathbf{P}_x(\tau_C < \infty) = 1$, $x \in E$. Let $\Delta = \{\mathfrak{z}_0 \in C\}$. Then $0 < \mathbf{P}_\pi(\Delta) < \infty$ and, by Proposition 16, the transformation T_{τ_C} preserves the measure $\mathbf{P}_\pi(\cdot|\Delta) = \mathbf{P}_\rho(\cdot)$, where $\rho(A) = \dfrac{1}{\pi(C)}\pi(AC)$. In other words, the Markovian chain $\mathfrak{y}_n = \mathfrak{z}_{\tau_n}$ $(n = 1, 2, \ldots)$, imbedded in the sequence $\mathfrak{z}_n$ $(n = 1, 2, \ldots)$, at times τ_n at which it attains the set C, has an invariant probability distribution ρ. The recurrence, with respect to μ, together with the previous statement, insures that the chain $\mathfrak{h}_n$ $(n = 1, 2, \ldots)$ is ergodic, the measure ρ is unique and π can be expressed in terms of ρ by (59). This demonstrates the required uniqueness which proves the Theorem.

34. One should not mix the notion of Harris recurrence with the existing notion of topological recurrence. A Markovian chain $\mathfrak{z}_n$ $(n = 0, 1, 2, \ldots)$, is called *topologically recurrent* if the measurable structure on its phase space is generated by some topology and any open set of the topology is attainable from any initial state with unit probability.

The difference between these notions can be observed from the example of a random walk. Namely, let $\mathfrak{z}_n = \mathfrak{z}_0 + \xi_1 + \ldots + \xi_n$, where the numeric random variables $\xi_1, \xi_2, \ldots$ are jointly independent, equally distributed, and do not depend on $\mathfrak{z}_0$. If their joint distribution, F, is not concentrated at points $0, \pm\delta, \pm 2\delta, \ldots$ for any $\delta > 0$, then a unique invariant σ-finite measure for the chain $\mathfrak{z}_n$ $(n = 0, 1, \ldots)$ is the Lebesgue measure.

If also $\int_{-\infty}^{\infty} |x| F(dx) < \infty$ then the random walk is topologically recurrent with respect to the natural topology of the real line if and only if $\int_{-\infty}^{\infty} x F(dx) = 0$. However, a topologically recurrent random walk is Harris recurrent (with respect to the Lebesgue measure) if and only if the convolution of F with itself n times is not singular with respect to the Lebesgue measure for any $n \geq 1$. This condition is obviously necessary, and a proof of sufficiency would make a good exercise for the interested reader.

35. *A random walk along a circle* of unit circumference (Example 25) need not necessarily be an *ergodic Markovian chain.* Namely, if all the convolutions, F^{*n} $(n = 1, 2, \ldots)$, of the distribution, F, are singular with itself with respect to the Lebesgue measure, then for every point, x, from the unit interval $[0, 1)$ there exists a Borel function g, for which;

$$\mathbf{P}_x \left(\lim_{n\to\infty} \frac{1}{n} \sum_{k=0}^{n-1} g(\zeta_k) = \int_0^1 g(x)dx \right) = 0. \tag{61}$$

For a given $x \in [0, 1)$, this function can be taken to be;

$$g(y) = i_A(y - x(\operatorname{mod} 1)), \qquad y \in [0, 1),$$

where the Borel subset A of the interval $[0, 1)$ has the Lebesgue measure equal to zero and $\mathbf{P}_0(\zeta_n \in A) = 1$ $(n = 1, 2, \ldots)$ (existence of such a set A follows from the singularity of F^{*n} $(n = 1, 2, \ldots)$). To check (61), it is sufficient to notice that;

$$\mathbf{P}_x(\zeta_n - x(\operatorname{mod} 1) \in A) = \mathbf{P}_0(\zeta_n \in A) = 1, \qquad n = 1, 2, \ldots,$$

and, consequently, $\lim_{n\to\infty} \frac{1}{n} \sum_{k=1}^{n} g(\zeta_k) = 1$ is almost certainly true with respect to $\mathbf{P}_x$, but at the same time, $\int_0^1 g(x)dx = 0$.

2.2 Sometimes Markovian processes

36. A measurable random process $\mathfrak{z}_t$ $(t \geq 0)$ with a phase space $(E, \mathcal{A})$ and the set of trajectories closed with respect to translations, is called *Markovian at time* $\mathfrak{m}$ if the conditional probabilities $\mathbf{P}_x(\cdot) = \mathbf{P}(\cdot|\mathfrak{z}_0 = x)$ are regular on the σ-algebra $\mathcal{M}$ generated by the process $\mathfrak{z}_t$, $t \geq 0$, and:

$$\mathbf{P}_x(T_{\mathfrak{m}}^{-1}\Gamma|\mathcal{M}_{\mathfrak{m}}) = \mathbf{P}_{\mathfrak{z}_{\mathfrak{m}}}(\Gamma) \tag{62}$$

is almost certainly valid, with respect to $\mathbf{P}_x$, for all $x \in E$, $\Gamma \in \mathcal{M}$. Here, $T_{\mathfrak{m}}$ is the translation operator by $\mathfrak{m}$ of a trajectory of the process $\mathfrak{z}_t$ $(t \geq 0)$. The nonnegative random variable $\mathfrak{m}$ is a stopping time with respect to the flow of σ-algebras $\mathcal{M}_t$ $(t \geq 0)$, which is generated by the process $\mathfrak{z}_t$ $(t \geq 0)$ (i.e. $\{\mathfrak{m} \leq t\} \in \mathcal{M}$ for all $t \geq 0$). The σ-algebra $\mathcal{M}_{\mathfrak{m}}$ contains those and only those events Γ from $\mathcal{M}$, for which $\Gamma \cap \{\mathfrak{m} \leq t\} \in \mathcal{M}_t$ for all $t \geq 0$.

If the Markovian property (62) holds at a certain, but not given at the beginning, stopping time $\mathfrak{m}$, then the process $\mathfrak{z}_t$ $(t \geq 0)$ is called *Virtual Markovian.*

37. The general theory of Virtual Markovian processes can be noticeably simplified (which can be observed from the following lemma) if it is assumed that the set of trajectories of the process $\mathfrak{z}_t$ $(t \geq 0)$ is closed with respect to the stopping times. This implies that, for every one of its elements ω, the space of elementary events, Ω, contains, for every $s \geq 0$, an element ω' such that $\mathfrak{z}_t(\omega') = \mathfrak{z}_{s\wedge t}(\omega)$ for all $t \geq 0$.

Lemma. *If the set of trajectories of a process* $\mathfrak{z}_t$ $(t \geq 0)$*, which is Markovian at time* $\mathfrak{m}$*, is closed with respect to the stopping times, then equation (62) is preserved if* $\mathfrak{m}$ *is replaced with any of the times* $\mathfrak{m}_1 = \mathfrak{m}$, $\mathfrak{m}_k = \mathfrak{m} + T_{\mathfrak{m}}\mathfrak{m}_{k-1}$, $k = 2, 3, \ldots$.

Proof. The proof is based on the following form of the definition given in (62) which is apparently more general. Let $\eta_t = \eta(t, \omega)$ be a bounded measurable (with respect to $\mathcal{B}_+ \times \mathcal{M}$) function. Then, if $\xi_t = T_t\eta_t$;

$$\mathbf{P}_x(\xi_{\mathfrak{m}}|\mathcal{M}_{\mathfrak{m}}) = \psi(\mathfrak{m}, \mathfrak{z}_{\mathfrak{m}}) \tag{63}$$

and is almost certainly true with respect to P_x, where $\psi(t, x) = \mathbf{P}_x\eta_t$.

Indeed, if:

$$\eta(t, \omega) = \begin{cases} 1, & \text{if } t < u,\ \omega \in \Gamma, \\ 0, & \text{otherwise} \end{cases} \tag{64}$$

for some $u > 0$ ($\Gamma \in \mathcal{M}$) then (63) becomes:

$$\mathbf{P}_x(\mathfrak{m} < u, T_{\mathfrak{m}}^{-1}\Gamma | \mathcal{M}_{\mathfrak{m}}) = \mathbf{P}_{\mathfrak{z}_{\mathfrak{m}}}(\Gamma)$$

is almost certainly true with respect to P_x, $\mathfrak{m} < u$. However, this evidently follows from (62) because the variable $\mathfrak{m}$ is measurable with respect to $\mathcal{M}_{\mathfrak{m}}$. The general case is proved by making an approximation to an arbitrary measurable (with respect to $\mathcal{B}_+ \times \mathcal{M}$) function by limits of the linear combinations of functions of the type given in (64).

If the set of trajectories of the process $\mathfrak{z}_t$ ($t \geq 0$) is closed with respect to the stopping time, then the σ-algebra $\mathcal{M}_{\mathfrak{m}}$ is generated by the variables $\mathfrak{z}_{t\wedge \mathfrak{m}}$. Indeed, by the axiom of choice, for every $u > 0$ there exists such a mapping $S_u : \Omega \to \Omega$ that $\mathfrak{z}_t(S_u\omega) = \mathfrak{z}_{t\wedge u}(\omega)$ for all $t \geq 0$. Any such mapping is a measurable mapping of the measurable space $(\Omega, \mathcal{M})$ into itself because the inverse image of the set $\{\omega : \mathfrak{z}_t(\omega) \in B\}$ under the mapping S_u, where $B \in \mathcal{A}$, is the set $\{\omega : \mathfrak{z}_{t\wedge u}(\omega) \in B\}$, which belongs to $\mathcal{M}_u$. Hence, $S_u^{-1}(\mathcal{M}) \subset \mathcal{M}_u$ and, if $\Gamma \in \mathcal{M}_u$, then $S_u^{-1}\Gamma = \Gamma$. Consequently, all sets from $\mathcal{M}_u$ and only these are invariant under the mapping S_u. Hence, for a measurable, with respect to $\mathcal{M}$, variable ξ to be measurable with respect to $\mathcal{M}_u$, it is necessary and sufficient that $\xi(\omega) = \xi(S_u\omega)$, or $\xi(\omega) = \xi(\omega')$ if $\mathfrak{z}_t(\omega) = \mathfrak{z}_t(\omega')$ for $t \leq u$. Because, for a measurable (with respect to $\mathcal{M}$) variable, ξ, its measurability, with respect to $\mathcal{M}_{\mathfrak{m}}$, implies that;

$$\xi_u = \begin{cases} \xi, & \text{if } \mathfrak{m} \leq u, \\ 0, & \text{if } \mathfrak{m} > u, \end{cases}$$

is measurable with respect to $\mathcal{M}_u$ for any $u \geq 0$. That is why the following condition is necessary and sufficient for a measurable, with respect to $\mathcal{M}$, variable, ξ, to be measurable with respect to $\mathcal{M}_{\mathfrak{m}}$:

$$\xi(\omega) = \xi(\omega'), \qquad \text{if } \mathfrak{z}_t(\omega) = \mathfrak{z}_t(\omega') \text{ for } t \leq \mathfrak{m}(\omega). \tag{65}$$

This is a criterion for measurability with respect to $\mathcal{M}_{\mathfrak{m}}$, from which, in particular, it follows that $\mathfrak{m}_k$ is a stopping time with respect to $\mathcal{M}_t$ ($t \geq 0$) for any natural k.

Proof. This is proved by induction. For $k = 1$, this is true by definition of $\mathfrak{m}_k$. Let $\mathfrak{m}_{k-1}$ be a stopping time, i.e. $\{\mathfrak{m}_{k-1} \leq u\} \in \mathcal{M}_u$ for all $u \geq 0$. Then, if $\mathfrak{z}_t(\omega) = \mathfrak{z}_t(\omega')$ for $t \leq u$ and $\mathfrak{m}_k(\omega) \leq u$, then $\mathfrak{m}_{k-1}(\omega) = \mathfrak{m}_{k-1}(\omega') \leq u$. Thus, $\mathfrak{m}_k(\omega') = \mathfrak{m}_{k-1}(\omega') + \mathfrak{m}(T_{\mathfrak{m}_{k-1}(\omega')}\omega') = v + \mathfrak{m}(T_v\omega')$, where $v = \mathfrak{m}_{k-1}(\omega) = \mathfrak{m}_{k-1}(\omega')$. This implies that $\mathfrak{m}(T_v\omega') = \mathfrak{m}(T_v\omega)$ and, consequently, $\mathfrak{m}_k(\omega) = \mathfrak{m}_k(\omega')$. Hence, the inequality $\mathfrak{m}_k(\omega) \leq u$ implies that $\mathfrak{m}_k(\omega') \leq u$. By symmetry, it follows that $\mathfrak{m}_k(\omega) \leq u$ if and only if $\mathfrak{m}_k(\omega') \leq u$, i.e. the event $\{\mathfrak{m}_k \leq u\}$ is measurable with respect to the σ-algebra $\mathcal{M}_u$, or $\mathfrak{m}_k$ is a stopping time with respect to $\mathcal{M}_u$ ($u \geq 0$).

Let $\mathcal{G}$ be a σ-algebra generated by the variables $\mathfrak{z}_{t\wedge \mathfrak{m}}$ ($t \geq 0$). Because the process $\mathfrak{z}_t$ ($t \geq 0$) is measurable, the mapping $S_{\mathfrak{m}}\omega = S_{\mathfrak{m}(\omega)}\omega$ is measurable as a mapping of

$(\Omega, \mathcal{M})$ into itself, and $\mathcal{G}$ is a σ-algebra of invariant, with respect to $S_{\mathfrak{m}}$, events from $\mathcal{M}$. Thus a measurable (with respect to $\mathcal{M}$) variable, ξ, is measurable with respect to $\mathcal{G}$, if and only if $\xi(\omega) = \xi(S_{\mathfrak{m}}\omega)$, which in its turn is equivalent to the condition $\xi(\omega)\xi(\omega')$ if $\mathfrak{z}_t(\omega) = \mathfrak{z}_t(\omega')$ for $t \le \mathfrak{m}(\omega)$ and so $\mathcal{G} = \mathcal{M}_{\mathfrak{m}}$.

If induction is used with (63), then the Markovian property given in (62) is preserved when $\mathfrak{m}$ is replaced by $\mathfrak{m}_k$. Let the lemma be proved for the times $\mathfrak{m}_1, \ldots <$ $\mathfrak{m}_{k-1}$. To complete the proof, the following verification should be undertaken:

$$\begin{aligned} &\mathbf{P}_x(\mathfrak{z}_{t_1} \in B_1, \ldots, \mathfrak{z}_{t_n} \in B_n, t_n \le \mathfrak{m}_k, \mathfrak{z}_{\mathfrak{m}_k} \in A, T_{\mathfrak{m}_k}^{-1}\Gamma) \\ &\quad = \mathbf{P}_x(\mathfrak{z}_{t_1} \in B_1, \ldots, \mathfrak{z}_{t_n} \in B_n t_n \le \mathfrak{m}_k, \mathfrak{z}_{\mathfrak{m}_k} \in A, \mathbf{P}_{\mathfrak{z}_{\mathfrak{m}_k}}(\Gamma)), \end{aligned} \tag{66}$$

where $x \in E$, $B_1, \ldots, B_n, A \in \mathcal{A}$, $0 < t_1 < \ldots < t_n$, $n = 1, 2, \ldots$, $\Gamma \in \mathcal{M}$. The event $\{\mathfrak{z}_{t_1} \in B_1, \ldots, \mathfrak{z}_{t_n} \in B_n, t_n \le \mathfrak{m}_k, \mathfrak{z}_{\mathfrak{m}_k} \in A\}$ can be represented as a union of the following nonintersecting events:

$$\begin{gathered} \{\mathfrak{z}_{t_1} \in B_1, \ldots, \mathfrak{z}_{t_1} \in B_n, t_n \le \mathfrak{m}, \mathfrak{z}_{\mathfrak{m}_k} \in A\} \cup \\ \cup \textstyle\bigcup_{j=1}^n \{\mathfrak{z}_{t_1} \in B_1, \ldots, \mathfrak{z}_{t_{j-1}} \in B_{j-1}, t_{j-1} \le \mathfrak{m}\} \cap \Lambda_{\mathfrak{m}}^j, \end{gathered}$$

where $\Lambda_s^j = T_s^{-1}\Delta_s^j$, $\Delta_s^j = \{\mathfrak{z}_{t_j - s} \in B_j, \ldots, \mathfrak{z}_{t_n - s} \in B_n, 0 < t_j - s, t_n - s \le$ $\mathfrak{m}_{k-1}, \mathfrak{z}_{\mathfrak{m}_{k-1}} \in A\}$. By (62) the following is obtained:

$$\begin{aligned} &\mathbf{P}_x(\mathfrak{z}_{t_1} \in B_1, \ldots, \mathfrak{z}_{t_{j-1}} \in B_{j-1}, t_{j-1} \le \mathfrak{m}, \Lambda_{\mathfrak{m}}^l, T_{\mathfrak{m}_k}^{-1}\Gamma) \\ &\quad = \mathbf{P}_x(\mathfrak{z}_{t_1} \in B_1, \ldots, \mathfrak{z}_{t_{j-1}} \in B_{j-1}, t_{j-1} \le \mathfrak{m}, \psi_j(\mathfrak{m}, \mathfrak{z}_{\mathfrak{m}})), \end{aligned} \tag{67}$$

where $\psi_j(s, x) = \mathbf{P}_x(\Delta_s^j T_{\mathfrak{m}_{k-1}}^{-1}(\Gamma))$. However, by the induction hypothesis, $\psi_j(s, x) =$ $\mathbf{P}_x(\Delta_s^j, \mathbf{P}_{\mathfrak{z}_{\mathfrak{m}_{k-1}}}(\Gamma))$. If formula (62) is applied to the right hand side of (67), then the left hand side of (67) can be written as:

$$\mathbf{P}_x(\mathfrak{z}_{t_1} \in B_1, \ldots, \mathfrak{z}_{t_{j-1}} \in B_{j-1}, t_{j-1} \le \mathfrak{m}, \Lambda_{\mathfrak{m}}^l, \mathbf{P}_{\mathfrak{z}_{\mathfrak{m}_k}}(\Gamma)). \tag{68}$$

Similarly,

$$\begin{aligned} &\mathbf{P}_x(\mathfrak{z}_{t_1} \in B_1, \ldots < \mathfrak{z}_{t_n} \in B_n, t_n \le \mathfrak{m}, \mathfrak{z}_{\mathfrak{m}_k} \in A, T_{\mathfrak{m}_k}^{-1}\Gamma) \\ &\quad = \mathbf{P}_x(\mathfrak{z}_{t_1} \in B_1, \ldots, \mathfrak{z}_{t_n} \in B_n, t_n \le \mathfrak{m}, \mathbf{P}_{\mathfrak{z}_{\mathfrak{m}}}(\mathfrak{z}_{\mathfrak{m}_{k-1}} \in A, T_{\mathfrak{m}_{k-1}}^{-1}\Gamma)) \\ &\quad = \mathbf{P}_x(\mathfrak{z}_{t_1} \in B_1, \ldots, \mathfrak{z}_{t_n} \in B_n, t_n \le \mathfrak{m}, \mathbf{P}_{\mathfrak{z}_{\mathfrak{m}}}(\mathbf{P}_{\mathfrak{z}_{\mathfrak{m}_{k-1}}}(\mathfrak{z}_0 \in A, \Gamma))) \\ &\quad = \mathbf{P}_x(\mathfrak{z}_{t_1} \in B_1, \ldots, \mathfrak{z}_{t_n} \in B_n, t_n \le \mathfrak{m}, \mathbf{P}_{\mathfrak{z}_{\mathfrak{m}}}(\mathfrak{z}_{\mathfrak{m}_{k-1}} \in A, \mathbf{P}_{\mathfrak{z}_{\mathfrak{m}_{k-1}}}(\Gamma))) \\ &\quad = \mathbf{P}_x(\mathfrak{z}_{t_1} \in B_1, \ldots, \mathfrak{z}_{t_n} \in B_n, t_n \le \mathfrak{m}, \mathfrak{z}_{\mathfrak{m}_k} \in A, \mathbf{P}_{\mathfrak{z}_{\mathfrak{m}_k}}(\Gamma)). \end{aligned}$$

By adding to this, the expressions given in (68), summed from $j = 1$ to n, equation (66) is obtained.

38. This lemma, in particular, implies that the sequence $\mathfrak{z}_{\mathfrak{m}_k}$ $(k = 1, 2, \ldots)$, imbedded into the process $\mathfrak{z}_t$ $(t \geq 0)$, is a homogeneous Markovian chain.

Moreover, it becomes clear from Lemma 37 that, if τ is an integer valued stopping time with respect to the flow of the σ-algebras, $\mathcal{M}_{\mathfrak{m}_k}$ $(k = 1, 2, \ldots)$ (i.e. $\{\tau \in k\} \in \mathcal{M}_{\mathfrak{m}_k}$, $k = 1, 2, \ldots$), then, in equalities (62) and (63), $\mathfrak{m}$ can be replaced by $\mathfrak{m}_\tau$. The event $\{\mathfrak{m}_\tau \leq u\}$ is equal to the union of the events $\{\tau = k, \mathfrak{m}_k \leq u\}$, each of which belongs to $\mathcal{M}_{\mathfrak{m}_k}$ because $\{\tau = k\} \in \mathcal{M}_{\mathfrak{m}_k}$ and any event from $\mathcal{M}_u$, being an intersection with $\{\mathfrak{m}_k \leq u\}$, gives, by definition, an event from $\mathcal{M}_u$ such that $\mathfrak{m}_\tau$ will be a stopping time with respect to $\mathcal{M}_t$ $(t \geq 0)$.

39. It is useful to remember that the criterion given in (65) of the measurability with respect to $\mathcal{M}_{\mathfrak{m}}$ formulated in the proof of Lemma 37, is not valid without the assumption that the set of trajectories is closed with respect to the stopping times. Indeed, let the set of time parameters of a process consist of two points i.e. zero and unity, and let the set Ω of elementary events be the interval on the number line, $[0, 1]$. Also, let $\mathfrak{m} \equiv 0$, $\mathfrak{z}_0(\omega) = \omega$ and $\mathfrak{z}_1(\omega) = i_D(\omega)$, where D is a non-Borel subset of the interval $[0, 1]$. Then, $\mathcal{M}_0$ is a σ-algebra of the Borel subsets of $[0, 1]$, and the σ-algebra M_1 is generated by all sets from $\mathcal{M}_0$ and the set D. If $\mathfrak{z}_0(\omega) = \mathfrak{z}_0(\omega')$, then $\mathfrak{z}_1(\omega) = \mathfrak{z}_1(\omega')$, which contradicts the fact that the variable $\mathfrak{z}_1$ is not measurable with respect to $\mathcal{M}_0$.

40. Theorem. *Let the set of trajectories of a process, $\mathfrak{z}_t$, which is Markovian at time $\mathfrak{m}$ $(t \geq 0)$, with a phase space, $(E, \mathcal{A})$, be closed with respect to the stopping times, and the σ-algebra, $\mathcal{A}$, be countably generated. Then, if the imbedded Markovian chain $\mathfrak{z}_{\mathfrak{m}_k}$ $(k = 1, 2, \ldots)$ is recurrent with an invariant measure π and:*

$$\mathbf{P}_\pi(\mathfrak{m} = 0) = 0, \tag{69}$$

then;

$$\lim_{t \to \infty} \frac{\int_0^t T_u \xi du}{\int_0^t T_u \eta du} = \frac{\mathbf{P}_\pi \int_0^{\mathfrak{m}} T_t \xi dt}{\mathbf{P}_\pi \int_0^{\mathfrak{m}} T_t \eta dt} \tag{70}$$

is almost certainly true with respect to $\mathbf{P}_x$, where $x \in E$, $\mathbf{P}_\pi \int_0^{\mathfrak{m}} T_t(|\xi| + |\eta|)dt < \infty$, and $\mathbf{P}_\pi \int_0^{\mathfrak{m}} T_t \eta dt \neq 0$.

Proof. The Markovian property at time $\mathfrak{m}$ and the invariance of the σ-finite measure, π, implies the invariance, with respect to $T_{\mathfrak{m}}$, of the σ-finite measure, $\mathbf{P}_\pi$, on $\mathcal{M}$. Indeed, $\mathbf{P}_\pi T_{\mathfrak{m}} \xi = \mathbf{P}_\pi \mathbf{P}_{\mathfrak{z}_{\mathfrak{m}}} \xi = \mathbf{P}_\pi g(\mathfrak{z}_{\mathfrak{m}})$, where $g(x) = \mathbf{P}_x \xi$. But $\mathbf{P}_\pi g(\mathfrak{z}_{\mathfrak{m}}) = \int_E \pi(dx) g(x) = \mathbf{P}_\pi \xi$.

41. Lemma. *Suppose that conditions of the theorem are valid and let a measurable, with respect to $\mathcal{M}_{\mathfrak{m}}$, variable, ζ, be positive almost everywhere with respect to $\mathbf{P}_\pi$.*

Then;

$$\sum_{k=1}^{\infty} T_{\mathfrak{m}_k}\zeta = \infty \tag{71}$$

is true almost everywhere with respect to $\mathbf{P}_\pi$.

Proof. Without loss of generality, it can be assumed that $0 < \zeta \le c$ is true almost everywhere with respect to $\mathbf{P}_\pi$ and $\pi\{x : \mathbf{P}_x\zeta > 1/c\} > 0$. Suppose that (71) is not valid. Then, there exists a time, t, such that $t > 0$ and;

$$\mathbf{P}_\pi(\lim_{k\to\infty} \sigma_k \le t) > 0, \tag{72}$$

where $\sigma_n = \sum_{k=0}^{n} T_{\mathfrak{m}_k}\zeta$, $\mathfrak{m}_0 = 0$.

$$\begin{aligned}
&\mathbf{P}_x(\sigma_k, \sigma_k \le t) - \mathbf{P}_x(\zeta, \zeta \le t) \\
&\quad = \sum_{j=1}^{k}[\mathbf{P}_x(\sigma_j, \sigma_j \le t) - \mathbf{P}_x(\sigma_{j-1}, \sigma_{j-1} \le t)] \\
&\quad = \sum_{j=1}^{k}[\mathbf{P}_x(T_{\mathfrak{m}_j}\zeta, \sigma_{j-1} \le t) - \mathbf{P}_x(\sigma_j, \sigma_{j-1} \le t < \sigma_j)].
\end{aligned}$$

Because $\sigma_j \le t + c$ if $\sigma_{j-1} \le t$, then;

$$\mathbf{P}_x(\sigma_j, \sigma_{j-1} \le t < \sigma_j) \le (t+c)\mathbf{P}_x(\sigma_{j-1} \le t < \sigma_j)$$

and:

$$\sum_{j=1}^{k} \mathbf{P}_x(\sigma_j, \sigma_{j-1} \le t < \sigma_j) \le (t+c)\mathbf{P}_x(\sigma_k > t).$$

By using the measurability with respect to $\mathcal{M}_{\mathfrak{m}}$ criterion from the proof of Lemma 37, the event $\{\sigma_{j-1} \le t\}$ is measurable with respect to the σ-algebra $\mathcal{M}_{\mathfrak{m}_j}$. This is proved by induction. It is true for $j = 1$. Let the variable ω_{j-1} be measurable with respect to $\mathcal{M}_{\mathfrak{m}_j}$, and $\mathfrak{z}_t(\omega) = \mathfrak{z}_t(\omega')$ for $t \ge \mathfrak{m}_{j+1}(\omega)$. Then $\mathfrak{m}_j(\omega) = \mathfrak{m}_j(\omega')$, $\mathfrak{m}_{j+1}(\omega) = \mathfrak{m}_{j+1}(\omega)$, $\sigma_{j-1}(\omega) = \sigma_{j-1}(\omega')$ and, consequently;

$$\sigma_j(\omega) = \sigma_{j-1}(\omega) + \zeta(T_{\mathfrak{m}_j(\omega')}\omega).$$

However, $\mathfrak{z}_t(T_{\mathfrak{m}_j(\omega')}\omega) = \mathfrak{z}_t(T_{(\mathfrak{m}_j\omega')}\omega')$ for $t \le \mathfrak{m}_{j+1}(\omega') - \mathfrak{m}_j(\omega') = \mathfrak{m}(T_{\mathfrak{m}_j(\omega')}\omega)$, and since the variable, ζ, is measurable with respect to $\mathcal{M}_{\mathfrak{m}}$, the following is obtained:

$$\zeta(T_{\mathfrak{m}_j(\omega')}\omega) = \zeta(T_{\mathfrak{m}_j(\omega')}\omega'),$$

which implies that $\sigma_j(\omega)$ and $\sigma_j(\omega')$ coincide.

Because at the time, $\mathfrak{m}_j$, the process has the Markovian property;

$$\mathbf{P}_x(T_{\mathfrak{m}_j}\zeta, \sigma_{j-1} \le t) = \mathbf{P}_x(\sigma_{j-1} \le t, \mathbf{P}_{\mathfrak{z}_{\mathfrak{m}_j}}\zeta)$$

$$\geq \frac{1}{c}\mathbf{P}_x(\sigma_{j-1} \leq t, \mathfrak{z}_{\mathfrak{m}_j} \in A) \geq \frac{1}{c}\mathbf{P}(\mathfrak{z}_{\mathfrak{m}_j} \in A, \lim_{k\to\infty} \sigma_k \leq t),$$

where $A = \{x : \mathbf{P}_x\zeta > 1/c\}$. This implies:

$$\begin{aligned} t\mathbf{P}_x(\sigma_k \leq t) &\geq \mathbf{P}_x(\sigma_k, \sigma_k \leq t) \\ &\geq -(t+c)\mathbf{P}_x(\sigma_k > t) + \frac{1}{c}\sum_{j=1}^{k} \mathbf{P}_x(\mathfrak{z}_{\mathfrak{m}_k} \in A, \lim_{r\to\infty} \sigma_r \leq t) \end{aligned}$$

and, consequently;

$$\sum_{j=1}^{\infty} \mathbf{P}_x(\mathfrak{z}_{\mathfrak{m}_j} \in A, \lim_{r\to\infty} \sigma_r \leq t) \leq c(t+c).$$

This inequality yields the convergence of the series

$$\sum_{j=1}^{k} \mathbf{P}_x(\mathfrak{z}_{\mathfrak{m}_j} \in A, \lim_{r\to\infty} \sigma_r \leq t),$$

and the Borel-Cantelli Theorem shows that there are only a finite number of events:

$$\{\mathfrak{z}_{\mathfrak{m}_j} \in A, \int_{r\to\infty} \sigma_r \leq t\}, \qquad j = 1, 2, \ldots \tag{73}$$

which almost certainly take place with respect to $\mathbf{P}_x$.

However, $\pi(A) > 0$ and the Markovian chain $\mathfrak{z}_{\mathfrak{m}_j}$ $(j = 1, 2, \ldots)$ is recurrent. Thus, $\mathbf{P}_x(\mathfrak{z}_{\mathfrak{m}_j} \in A$ for infinitely many j $) = 1$. Hence, the probability, $\mathbf{P}_x$, that infinitely many events, described in (73), will occur is equal to:

$$\mathbf{P}_x(\lim_{r\to\infty} \sigma_r \leq t),$$

and, as (72) demonstrates, this probability is positive for a positive, with respect to π, set of x.

This contradiction proves the lemma, which, by (69) and the equation $\mathfrak{m}_k = \sum_{j=0}^{k-1} T_{\mathfrak{m}_j}\mathfrak{m}$, implies that the sequence $\mathfrak{m}_k$ $(k = 1, 2, \ldots)$ is unbounded almost everywhere, with respect to $\mathbf{P}_\pi$.

Let the set X of number sequences, $x = (x_0, x_1, \ldots)$, and the σ-algebra, $\mathcal{C}$, of its subsets be defined in the same way as in the proof of Corollary 7. Let a measurable (with respect to $\mathcal{M}_{\mathfrak{m}}$) variable, ζ, be positive almost everywhere with respect to $\mathbf{P}_{\mathfrak{m}}$, and $\mathbf{P}_\pi\zeta < \infty$ (for example, $\zeta = \varphi(\mathfrak{z}_0)$, where $\varphi > 0$ and $\int_E \varphi(y)\pi(dy) < \infty$). The equation:

$$\begin{aligned} &\mu\{x \in X : x_0 < c_0, x_1 < c_1, x_2 < c_2, \ldots, x_{3k} < c_{3k}, x_{3k+1} < c_{3k+1}, x_{3k+2} < c_{3k+2}\} \\ &\quad = \mathbf{P}_\pi(\zeta_0 < c_0, \zeta_1 < c_1, \zeta_2 < c_2, \ldots, \zeta_{3k} < c_{3k}, \zeta_{3k+1} < c_{3k+1}, \zeta_{3k+2} < c_{3k+2}), \end{aligned}$$

where $k = 0, 1, 2, \ldots$, $\zeta_{3k} = T_{\mathfrak{m}_k}\zeta$, $\zeta_{3k+1} = \int_{\mathfrak{m}_k}^{\mathfrak{m}_{k+1}} T_t\xi dt$, $\zeta_{3k+2} = \int_{\mathfrak{m}_k}^{\mathfrak{m}_{k+1}} T_t\eta dt$ and $\mathfrak{m}_0 = 0$, defines, on $\mathcal{C}$, a σ-finite measure, μ, which is invariant with respect to the transformation, T_3, that translates a sequence, $x = (x_0, x_1, x_3, \ldots)$, to the right by three steps: $T_3x = (x_3, x_4, x_5, \ldots)$. The functions $f_0(x) = x_0$, $f_1(x) = x_1$ and $f_2(x) = x_2$ are integrable with respect to the measure, μ, and $f_0 > 0$ is true almost everywhere with respect to μ. By Lemma 41, the series $\sum_{k=1}^{\infty} T_{\mathfrak{m}_k}\zeta$ is divergent almost everywhere with respect to $\mathbf{P}_\pi$, which is equivalent to the divergence of the series $\sum_{k=1}^{\infty} f_0(T_3^k x)$ almost everywhere with respect to μ. By Theorem 4, the limits;

$$\lim_{n\to\infty} \frac{\sum_{k=1}^{n} f_1(T_3^k x)}{\sum_{k=1}^{n} f_0(T_3^k x)} = h_1(x), \qquad \lim_{n\to\infty} \frac{\sum_{k=1}^{n} f_2(T_3^k x)}{\sum_{k=1}^{n} f_0(T_3^k x)} = h_2(x)$$

exist almost everywhere with respect to μ and:

$$\int_X f_0(x)h_1(x)\mu(dx) = \int_X f_1(x)\mu(dx),$$
$$\int_X f_0(x)h_2(x)\mu(dx) = \int_X f_2(x)\mu(dx).$$

This is equivalent to the existence almost everywhere with respect to $\mathbf{P}_\pi$, of the limits;

$$\xi^* = \lim_{k\to\infty} \frac{\int_0^{\mathfrak{m}_k} T_u\xi du}{\sum_{j=0}^{k-1} T_{\mathfrak{m}_j}\zeta}, \qquad \eta^* = \lim_{k\to\infty} \frac{\int_0^{\mathfrak{m}_k} T_u\eta du}{\sum_{j=0}^{k-1} T_{\mathfrak{m}_j}\zeta}$$

and:

$$\mathbf{P}_\pi\xi^*\zeta = \mathbf{P}_\pi \int_0^{\mathfrak{m}} T_u\xi du, \qquad \mathbf{P}_\pi\eta^*\zeta = \mathbf{P}_\pi \int_0^{\mathfrak{m}} T_u\eta du.$$

The limit variables, ξ^* and η^*, are invariant with respect to the mapping, $T_{\mathfrak{m}}$, and because the chain $\mathfrak{z}_{\mathfrak{m}_k}$ $(k = 1, 2, \ldots)$ is recurrent, they are degenerate almost everywhere with respect to $\mathbf{P}_\pi$. Indeed, since the sequence $\mathfrak{m}_k$ $(k = 1, 2, \ldots)$ is unbounded almost everywhere with respect to $\mathbf{P}_\pi$, the minimal σ-algebra, which contains the monotone sequence of σ-algebras, $\mathcal{M}_{\mathfrak{m}_k}$ $(k = 1, 2, \ldots)$, may differ from $\mathcal{M}$ only on the sets with a zero measure $\mathbf{P}_\pi$. Hence, if a measurable (with respect to $\mathcal{M}$) variable, η, is bounded and $T_{\mathfrak{m}}\eta = \eta$, then the set of those x from E for which;

$$\eta = \mathbf{P}_x(\eta|\mathcal{M}) = \lim_{k\to\infty} \mathbf{P}_x(\eta|\mathcal{M}_{\mathfrak{m}_k})$$
$$= \lim_{k\to\infty} \mathbf{P}_x(T_{\mathfrak{m}_k}\eta|\mathcal{M}_{\mathfrak{m}_k}) = \lim_{k\to\infty} \mathbf{P}_{\mathfrak{z}_{\mathfrak{m}_k}}\eta$$

is true almost everywhere with respect to $\mathbf{P}_x$, has the full measure π.

Because the Markovian chain $\mathfrak{z}_{\mathfrak{m}_k}$ $(k = 1, 2, \ldots)$ is recurrent, the limit $\lim_{k\to\infty} \mathbf{P}_{\mathfrak{z}_{\mathfrak{m}_k}}\eta$ is almost certainly degenerate with respect to $\mathbf{P}_x$, for almost all x, with respect to π, and this proves that the σ-algebra of invariant events with respect to $T_{\mathfrak{m}}$ is trivial with respect to the measure $\mathbf{P}_\pi$. Hence;

$$\xi^* = \frac{\mathbf{P}_\pi \int_0^{\mathfrak{m}} T_u\xi du}{\mathbf{P}_\pi\zeta}, \qquad \eta^* = \frac{\mathbf{P}_\pi \int_0^{\mathfrak{m}} T_u\eta du}{\mathbf{P}_\pi\zeta}$$

is true almost everywhere with respect to $\mathbf{P}_\pi$ and, consequently;

$$\lim_{k\to\infty} \frac{\int_0^{\mathfrak{m}_k} T_u\xi du}{\int_0^{\mathfrak{m}_k} T_u\eta du} = \frac{\mathbf{P}_\pi \int_0^{\mathfrak{m}} T_u\xi du}{\mathbf{P}_\pi \int_0^{\mathfrak{m}} T_u\eta du}$$

is true almost everywhere with respect to $\mathbf{P}_\pi$.

The following inequalities, which are valid for $\xi \geq 0$ and $\mathfrak{m}_k \leq t \leq \mathfrak{m}_{k+1}$;

$$\frac{\int_0^{\mathfrak{m}_k} T_u\xi du}{\sum_{j=0}^{k-1} T_{\mathfrak{m}_j}\zeta} \leq \frac{\int_0^t T_u\xi du}{\sum_{j=0}^{k-1} T_{\mathfrak{m}_j}\zeta} \leq \frac{\sum_{j=0}^{k} T_{\mathfrak{m}_j}\zeta}{\sum_{j=0}^{k-1} T_{\mathfrak{m}_j}\zeta} \frac{\int_0^{\mathfrak{m}_{k+1}} T_u\zeta du}{\sum_{j=0}^{k} T_{\mathfrak{m}_j}\zeta},$$

together with the relation;

$$\lim_{k\to\infty} \frac{\sum_{j=0}^{k} T_{\mathfrak{m}_j}\zeta}{\sum_{j=0}^{k-1} T_{\mathfrak{m}_j}\zeta} = \frac{\mathbf{P}_\pi T_{\mathfrak{m}}\zeta}{\mathbf{P}_\pi\zeta} = 1,$$

being true almost everywhere with respect to $\mathbf{P}_\pi$, yields the existence, almost everywhere with respect to $\mathbf{P}_\pi$, of the limit in the left hand side of (70).

Thus (70) is almost certainly valid with respect to $\mathbf{P}_\pi$. Let Γ denote the event such that the limit in the left hand side of (70) exists and equals the right hand side of (70). Let B be the set of those x from E for which $\mathbf{P}_x(\Gamma) = 1$. Then, $\pi(\rceil B) = 0$ and, consequently, $\mathbf{P}_x(\tau_B < \infty) = 1$ for all $x \in E$, where $\tau_B = \min\{k : \mathfrak{z}_{\mathfrak{m}_k} \in B\}$. So, by the invariance of Γ, the almost certain (with respect to $\mathbf{P}_x$) inclusion of $\mathfrak{z}_{\mathfrak{m}_{\tau_B}} \in B$ and the Markovian property of the process at the time $\mathfrak{m}_{\tau_B}$, then;

$$\mathbf{P}_x(\Gamma) = \mathbf{P}_x(T^{-1}_{\mathfrak{m}_{\tau_B}}\Gamma) = \mathbf{P}_x\mathbf{P}_{\mathfrak{z}_{\mathfrak{m}_{\tau_B}}}(\Gamma) = 1,$$

which proves the statement.

42. A typical example of a Virtual Markovian process is a *semi-Markovian process.* Such a process, which shall be denoted by $\mathfrak{z}_t$ $(t \geq 0)$, is Markovian at the time, $\mathfrak{m} = \inf\{t : \mathfrak{z}_t \neq \mathfrak{z}_0\}$, of the first jump and has step trajectories, i.e. the trajectories are right continuous and have left limits in the discrete topology, the phase space $(E, \mathcal{A})$. The σ-algebra, $\mathcal{A}$, contains all one-point sets and is generated by a countable number of its elements. Then, the σ-algebra, $\mathcal{M}$, which is generated by the process $\mathfrak{z}_t$ $(t \geq 0)$, is also countably generated because it coincides with the minimal σ-algebra, with respect to which, the variables $\mathfrak{z}_0, \mathfrak{m}_1, \mathfrak{z}_{\mathfrak{m}_1}, \mathfrak{m}_2, \mathfrak{z}_{\mathfrak{m}_2}, \ldots$ are measurable. Therefore, there exists a regular (with respect to both x and u) variant, $\mathbf{P}^\bullet_{x,u}$, of the conditional probabilities $\mathbf{P}^\bullet_{x,u}(\Gamma) = \mathbf{P}(\Gamma|\mathfrak{z}_0 = x, \mathfrak{m} = u)$ and $\mathbf{P}_x\mathbf{P}^\bullet_{x,\mathfrak{m}}(\Gamma) = \mathbf{P}_x(\Gamma)$, $\Gamma \in \mathcal{M}$.

Let the imbedded Markovian chain, $\mathfrak{z}_{\mathfrak{m}_k}$ $(k = 1, 2, \ldots)$, in the process, $\mathfrak{z}_t$ $(t \geq 0)$, be recurrent with an invariant measure, π, and let the σ-finite measure, $\pi^\bullet$, be defined on the product $(E \times R_+, \mathcal{A} \times \mathcal{B}_+)$ in the following way:

$$\pi^\bullet(A \times [0, s)) = \int_A \pi(dx)\mathbf{P}_x(\mathfrak{m} \wedge s), \qquad A \in \mathcal{A},\ s > 0.$$

Then, if the conditions of Theorem 40 are satisfied;

$$\lim_{t\to\infty} \frac{\int_0^t T_u \xi du}{\int_0^t T_u \eta du} = \frac{\mathbf{P}^\bullet_{\pi\bullet}\xi}{\mathbf{P}^\bullet_{\pi\bullet}\eta} \tag{74}$$

is almost certainly true with respect to $\mathbf{P}_x$, where $x \in E$ and $\mathbf{P}^\bullet_{\pi\bullet} = \int_{E\times R_+} \pi^\bullet(dx \times du)\mathbf{P}^\bullet_{x,u}$.

It is sufficient to verify formula (74) in the case of $\xi = f(\mathfrak{z}_0, \mathfrak{m})T_\mathfrak{m}\eta$ because the whole σ-algebra $\mathcal{M}$ is generated by linear combinations of such variables, ξ. The function, f, is measurable with respect to $\mathcal{A}\times\mathcal{B}_+$, and the variable, η, is measurable with respect to $\mathcal{M}$. By Theorem 40, formula (74) is equivalent to the equation;

$$\mathbf{P}_\pi \int_0^\mathfrak{m} T_u \xi du = \mathbf{P}^\bullet_{\pi\bullet}\xi \tag{75}$$

or, for ξ of the type given above:

$$\mathbf{P}_\pi \int_0^\mathfrak{m} T_u[f(\mathfrak{z}_0, \mathfrak{m})T_\mathfrak{m}\eta]du = \int_{E\times R_+} \pi^\bullet(dx \times du) f(x,u)\mathbf{P}_{x,u}T_u\eta. \tag{75'}$$

Because;

$$\{u < \mathfrak{m}\} \subset \{T_u\mathfrak{m} = \mathfrak{m} - u\} \subset \{T_uT_\mathfrak{m}\eta = T_\mathfrak{m}\eta\},$$

then the left hand side of (75') is equal to:

$$\mathbf{P}_\pi\left[(T_\mathfrak{m}\eta)\int_0^\mathfrak{m} f(\mathfrak{z}_0, \mathfrak{m}-u)du\right] = \mathbf{P}_\pi\left[(T_\mathfrak{m}\eta)\int_0^\mathfrak{m} f(\mathfrak{z}_0, u)du\right]$$

$$\left(\text{because the integral } \int_0^\mathfrak{m} f(\mathfrak{z}_0,u)du \text{ is measurable with respect to } \mathcal{M}_\mathfrak{m}\right)$$

$$= \mathbf{P}_\pi\left[\int_0^\mathfrak{m} f(\mathfrak{z}_0,u)du\mathbf{P}_{\mathfrak{z}_\mathfrak{m}}\eta\right] = \int_{E\times R_+} \pi^\bullet(dx, du)f(x,u)\mathbf{P}^\bullet_{x,u}\mathbf{P}_{\mathfrak{z}_\mathfrak{m}}\eta.$$

On the other hand, equation (63) shows that;

$$\mathbf{P}^\bullet_{x,u}T_u\eta\eta = \mathbf{P}^\bullet_{x,u}T_\mathfrak{m}\eta = \mathbf{P}^\bullet_{x,u}\mathbf{P}_{\mathfrak{z}_\mathfrak{m}}\eta.$$

Which proves (75) and, together with (75), (74) is also proved.

Chapter 3

A random walk on the semiaxis

The major part of this chapter deals with the proof of the ergodic theorem for a semicontinuous random walk on the semi-axis which is understood to be a strictly Markovian (homogeneous) process in the phase space $(1, 2, \dots, d) \times [0, \infty)$ and is a process with independent increments in a random medium until it reaches the set of states $\{(i, 0) : i = 1, \dots, d\}$ for the first time.

To prove this theorem, ergodic theorems from the previous section shall be used. By choosing a particular time, $\mathfrak{m}$ (the time at which the "wandering" component reaches zero for the first time after it reaches a certain positive value), the problem can be formulated in terms of *boundary-value problems.*

First of all, the problem of all semicontinuous walks on the semi-axis described in relation to a given process with independent increments in a random medium has to be considered. This is equivalent to the definition of local characteristics of the process or, in other words, the definition of the terminology that will be used to formulate statements about this process.

3.1 Semicontinuous processes with independent increments in a random medium

43. A strictly Markovian process, (ν_t, ζ_t) $(t \geq 0)$, in the phase space, $\{1, \dots, d\} \times \mathbf{R}^1$, is called a *(homogeneous) process with independent increments in a random medium* if its conditional distribution, under the conditions $\nu_0 = i$ and $\zeta_0 = x$, coincides with the distribution of the process $(\nu_t, x + \zeta_t)$ $(t \geq 0)$ subject to the conditions $\nu_0 = i$ and $\zeta_0 = 0$.

44. Theorem. *If almost all trajectories of a process with independent increments in a random medium, (ν_t, ζ_t) $(t \geq 0)$, are right-continuous, have left limits, and*

$\mathbf{P}_{t,x}(\zeta_t - \zeta_{t-} \geq 0) = 1$, *then;*

$$\|\mathbf{P}_{i,0}(e^{s\zeta_t}, \nu_t = j)\|_{i,j=1}^d = \exp\{tA(s)\}, \tag{76}$$

where $\Re s \leq 0$ *and* $\mathbf{P}_{i,x}$ *are conditional probabilities for* $\nu_0 = i$ *and* $\zeta_0 = x$;

$$\begin{aligned} A(s) &= A_0 + \int_0^\infty L_0(dy)e^{sy} + A_1 s + A_2 s^2 \\ &+ \int_0^\infty L_1(dy)[e^{sy} - 1 - s(y \wedge 1)]. \end{aligned} \tag{77}$$

Here A_0, A_1, A_2 *and* $L_1(dy)$ *are diagonal matrices with the corresponding elements* a_i^0, a_i^1, a_i^2, $l_i(dy)$, $a_i^2 \geq 0$, $l_i(dy) \geq 0$, $l_i(0) = 0$ *and* $\int_0^\infty (t^2 \wedge 1) l_i(dy) < \infty$. *The matrix* $L_0(dy)$ *has all the diagonal elements equal to zero, its off-diagonal elements,* $l_{ij}(dy)$, *are nonnegative and:*

$$a_i^0 + \sum_{i \neq j} l_{ij}[0, \infty) = 0. \tag{78}$$

Proof. It is sufficient to establish the existence of the mean values in the left hand side of (76) for $s < 0$ in the case where $t > 0$ because, if $P_{i,0}e^{s\zeta_t} < \infty$ when $0 < t < t_0$ ($i = 1, \ldots, d$) and where t_0 can depend on s, then;

$$\mathbf{P}_{i,0}e^{s\zeta_t} = \mathbf{P}_{i,0}\mathbf{P}_{\nu_{t_0},\zeta_{t_0}} e^{s\zeta_t} = \mathbf{P}_{t,0}[e^{s\zeta_{t_0}} \mathbf{P}_{\nu_{t_0},0} e^{s\zeta_{t-t_0}}] < \infty$$

for $0 < t - t_0$ or $0 < t < 2t_0$.

Let $p_t(x) = \max_i \mathbf{P}_{t,0}(\inf_{0 \leq u \leq t} \zeta_u < x) = \max_i \mathbf{P}_{i,0}(\theta(x) < t)$, where $\theta(x) = \inf\{t : \zeta_t \leq x\}$. Then, $\lim_{t \to 0} p_t(x) = 0$ and $\mathbf{P}_{i,0}(\zeta_{\theta(x)} = x) = 1$ for $x < 0$ due to the continuous nature and the fact that there are no negative jumps.

If $x < 0$ and $y < 0$, then;

$$\begin{aligned} \mathbf{P}_{t,0}(\zeta_t < x + y) &= \mathbf{P}_{i,0}(\zeta_t < x + y, \theta(x) < t) \\ &= \mathbf{P}_{i,0}(\zeta_t - \zeta_{\theta(x)} + \zeta_{\theta(x)} < x + y, \theta(x) < t) \\ &\leq \mathbf{P}_{i,0}(\inf_{\theta(x) \leq u \leq t} \zeta_u - \zeta_{\theta(x)} < y, \theta(x) < t) \\ &\leq \mathbf{P}_{i,0}(\inf_{0 \leq u \leq t} \zeta_{u+\theta(x)} - \zeta_{\theta(x)} < y, \theta(x) < t) \\ &= \sum_j \mathbf{P}_{i,0}(\theta(x) < t, \nu_{\theta(x)} = j)\mathbf{P}_{j,0}(\inf_{0 \leq u \leq t} \zeta_u < y) \\ &\leq p_t(x)p_t(y) \end{aligned}$$

and, consequently, $p_t(x + y) \leq p_t(x)p_t(y)$.

Let a positive number, t_0, such that $t_0 = t_0(s)$, be so small that; $p_t(-1) \leq e^{2s}$ for $t \leq t_0$ and $s < 0$. Then, $\mathbf{P}_{t,0}(\zeta_t < -n) \leq p_t(-n) \leq p_t(-n+1)p_t(-1) \leq e^{2ns}$ and;

$$\mathbf{P}_{i,0}(e^{s\zeta_t}, \zeta_t < 0) = \sum_{n=0}^\infty \mathbf{P}_{i,0}(e^{s\zeta_t}, -n - 1 \leq \zeta_t < -n)$$

$$\leq e^{-s}\sum_{n=0}^{\infty} e^{-sn}\mathbf{P}_{\imath,0}(\zeta_t < -n) < \infty.$$

Further;

$$\begin{aligned}\mathbf{P}_{\imath,0}(e^{s\zeta_{t+u}}, \nu_{t+u} = \jmath) &= \mathbf{P}_{\imath,0}\mathbf{P}_{\nu_t,\zeta_t}(e^{s\zeta_u}, \nu_u = \jmath) \\ &= \mathbf{P}_{\imath,0}[e^{s\zeta_t}\mathbf{P}_{\nu_t,0}(e^{s\zeta_u}, \nu_u = \jmath)] \\ &= \sum_k \mathbf{P}_{\imath,0}(e^{s\zeta_t}, \nu_t = k)\mathbf{P}_{k,0}(e^{s\zeta_u}, \nu_u = \jmath).\end{aligned}$$

So, if the matrix in the left hand side of (76) is denoted by $F(t,s)$, then $F(t+u,s) = F(t,s)F(u,s)$ and, because $F(0+,s) = E$, then $F(t,s) = \exp\{tA(s)\}$. To obtain the matrix, $A(s)$, in the analytical form given in (77), the following argument can be used.

If a natural number, n, is large enough so that the half-plane, $\{z : \Re z > n\}$, does not contain points given in the spectrum of the matrix $A(s)$, then, for a given $c < 0$, the $(\imath,\jmath)^{\text{th}}$ element of the matrix $\{nE - A(s)\}^{-1}$ is equal to:

$$\begin{aligned}\mathbf{P}_{\imath,0}\int_0^{\infty} e^{-nt}e^{s\zeta_t}t_{\{\nu_t=\jmath\}}dt &= \mathbf{P}_{\imath,0}\int_0^{\theta(c)} e^{-nt}e^{s\zeta_t}\imath_{\{\nu_t=\jmath\}}dt \\ &\quad +\mathbf{P}_{\imath,0}e^{-n\theta(c)}e^{s\zeta_{\theta(c}}\int_0^{\infty} e^{-nt}e^{s[\zeta_{\theta(c)+t}-\zeta_{\theta(c)}]}\imath_{\{\nu_t=\jmath\}}dt \\ &= \mathbf{P}_{\imath,0}\int_0^{\theta(c)} e^{-nt}e^{s\zeta_t}\imath_{\{\nu_t=\jmath\}} + e^{sc}\sum_k \mathbf{P}_{\imath,0}[e^{-n\theta(c)}, \nu_t = k] \\ &\quad \times\mathbf{P}_{k,0}\int_0^{\infty} e^{-nt}e^{s\zeta_t}\imath_{\{\nu_t=\jmath\}}dt. \qquad (79)\end{aligned}$$

Because $\mathbf{P}_{\imath,0}(\theta(c) > 0) = 1$, then $\mathbf{P}_{\imath,0}e^{-n\theta(c)} \to 0$ and, because $\theta(c)$ is monotone with respect to c, there exists a sequence, $c_n \uparrow 0$, such that $\mathbf{P}_{\imath,0}e^{-n\theta(c_n)} \to 0$ as $n \to \infty \to 0$. Moreover, because $\lim_{n\to\infty} n\{nE - A(s)\}^{-1} = E$, then the left hand side of (79) remains bounded if multiplied by n and then n is taken to ∞. Hence;

$$n\sum_k \mathbf{P}_{\imath,0}[e^{-n\theta(c_n)}, \nu_t = k]\mathbf{P}_{k,0}\int_0^{\infty} e^{-nt}e^{s\zeta_t}\imath_{\{\nu_t=\jmath\}}dt \underset{n\to\infty}{\to} 0.$$

Therefore;

$$\begin{aligned}nA(s)\{nE - A(s)\}^{-1} &= n(\{nE - A(s)\}^{-1} - E) \\ &= \int_{c_n}^{\infty} e^{sy}L^{(n)}(dy) - nE + o(1), \qquad (80)\end{aligned}$$

where the $(\imath,\jmath)^{\text{th}}$ element of the matrix $L^{(n)}[c_n, x)$ is;

$$nP_{\imath,0}\int_0^{\theta(c)} e^{-nt}\imath_{\{\zeta_t<x\}}\imath_{\{\nu_t=\jmath\}}dt.$$

The left hand side of (80) approaches $A(s)$ as n tends to ∞, i.e.

$$A(s) = \lim_{n\to\infty} \left\{ \int_{c_n}^{\infty} e^{sy} L^{(n)}(dy) - nE \right\}, \qquad s \le 0, \tag{81}$$

where the left hand side is continuous with respect to $s \le 0$. Therefore, the $(i,j)^{\text{th}}$ off-diagonal element of the matrix $A(s)$, being the limit of the corresponding elements of the matrices $\int_{c_n}^{\infty} e^{sy} L^{(n)}(dy)$, is equal to $\int_0^{\infty} e^{sy} l_{ij}(dy)$ with some finite measure, $l_{ij}(dy)$. It follows from the identity;

$$A(s) - 2A(s-1) + A(s-2) = \lim_{n\to\infty} \int_{c_n}^{\infty} e^{sy}(1-e^{-y})^2 L^{(n)}(dy),$$

that the diagonal elements of the matrix $(1-e^{-y})^2 L^{(n)}(dy)$ weakly converge as n tends to ∞, i.e. there exists finite measures, $l_i^{\bullet}(dy)$, on $[0,\infty)$, such that the limits of the diagonal elements of the matrices $\int_{c_n}^{\infty} f(y)(1-e^{-y})^2 L^{(n)}(dy)$ are equal to $\int_0^{\infty} f(y) l_i^{\bullet}(dy)$ $(i = 1,\dots,d)$ for any function, f, which is continuous and bounded on $[0,\infty)$.

If f is the function $f(y) = [e^{sy} - 1 - s(y \wedge 1)](1-e^{-y})^{-2}$, then the diagonal elements of the matrix;

$$\int_{c_n}^{\infty} [e^{sy} - 1 - s(y \wedge 1)] L^{(n)}(dy) \tag{82}$$

will converge, as n tends to ∞, to $\int_0^{\infty} [e^{sy} - 1 - s(y\wedge 1)] l_i^{\bullet}(dy)$ $(i = 1,\dots,d)$. However, the matrix given in (82) is equal to:

$$\left(\int_{c_n}^{\infty} e^{sy} L^{(n)}(dy) - nE \right) - (L^{(n)}[c_n,\infty) - nE) - s \int_{c_n}^{\infty} (y \wedge 1) L^{(n)}(dy).$$

This, together with (81), implies the existence of the limits, a_i^0 $(i = 1,\dots,d)$ and a_i^1 $(i = 1,\dots,d)$, of the diagonal elements of the matrices $L^{(n)}[c_n,\infty) - nE$ and $\int_{c_n}^{\infty} (y \wedge 1) L^{(n)}(dy)$.

To obtain (77), $a_i^2 = 2 l_i^{\bullet}\{0\}$ $(i = 1,\dots,d)$ and $l_i(dy) = (1-e^{-y})^{-2} l_i^{\bullet}(dy)$ $(y > 0)$.

Condition (78) is equivalent to the fact that the sum of the elements in any row of the matrix $\exp\{tA(0)\}$ equals unity.

45. The parameters A_0, A_1, A_2, $L_1(dy)$ and $L_0(dy)$ of this representation are uniquely determined from the process (ν_t, ζ_t) $(t \ge 0)$. Indeed, $A_2 = \lim_{s\to-\infty} s^2 A(s)$ and;

$$A(s) - 2A(s-1) + A(s-2) = \int_0^{\infty} e^{sy}(1-e^{-y})^{-2}\{L_0(dy) + L_1(dy)\},$$

so that $L(dy) = L_0(dy) + L_1(dy)$ is uniquely recovered from $A(s)$ $(s \le 0)$ and finally;

$$A_1 = -A(-1) + A_2 + A_0 + \int_0^{\infty} (e^{-y} - 1 + y \wedge 1) L_1(dy) + \int_0^{\infty} e^{-y} L_0(dy).$$

46. The converse statement of Theorem 44 is also true. For any matrix, $A(s)$, of the form given in (77), the equation given in (76) defines some process with independent increments in the random medium, (ν_t, ζ_t) $(t \geq 0)$, such that almost all trajectories of the process are right continuous, have left limits, and $\mathbf{P}_{i,x}(\zeta_t - \zeta_{t-} \geq 0) = 1$.

Further, one of the simplest ways to construct the process, (ν_t, ζ_t) $(t \geq 0)$, from the matrix, $A(s)$, is considered below.

Let ν_t $(t \geq 0)$ be a homogeneous Markovian process with the set of states $\{1, \ldots, d\}$ and let the transition probability matrix for a time, t, be $\exp\{tA(0)\}$, i.e.

$$\|\mathbf{P}_i(\nu_t = j)\|_{i,j=1}^{d} = e^{tA(0)},$$

where $\mathbf{P}_i$ is the conditional probability when $\nu_0 = i$.

Let also, for every i $(i = 1, \ldots, d)$, a process η_t^i $(t \geq 0)$ have independent increments, $\mathbf{P}(\eta_0^i = 0) = 1$ and;

$$\mathbf{P}e^{s\eta_t^i} = \exp\left\{t\left[a_i^1 s + a_i^2 s^2 + \int_0^\infty (e^{sy} - 1 - s(y \wedge 1)) l_i(dy)\right]\right\}. \tag{83}$$

The processes, η_t^i $(t \geq 0,\ i = 1, \ldots, d)$, are jointly independent and do not depend on the process ν_t $(t \geq 0)$. Trajectories of each of the processes given by η_t^i $(t \geq 0)$ are right continuous, have left limits, and do not have negative jumps. Trajectories of the process ν_t $(t \geq 0)$ are also right continuous (in the discrete topology).

Finally, let a collection, $\xi(i,j)$ (where $i, j = 1, \ldots, d$ and $i \neq j$), of nonnegative jointly independent random variables, which are also independent of the processes ν_t $(t \geq 0)$ and η_t^i $(t \geq 0,\ i = 1, \ldots, d)$, be given by the distributions;

$$\mathbf{P}(\xi(i,j) < x) = \frac{l_{ij}[0,x)}{l_{ij}[0,\infty)} \qquad \text{if} \quad l_{ij}[0,\infty) > 0,\ x > 0,$$

and:

$$\mathbf{P}(\xi(i,j) = 0) = 1 \qquad \text{if} \quad l_{ij}[0,\infty) = 0.$$

By using these processes and variables, the process ζ_t $(t \geq 0)$ can be constructed as follows. Let $\mathfrak{t}_1$, $\mathfrak{t}_2$, ... be a sequence of times at which the process ν_t $(t \geq 0)$ has jumps. Then;

$$\begin{aligned}
\zeta_t &= \eta_t^{\nu_0}, \qquad \text{if} \quad t \in [0, \mathfrak{t}_1), \\
\zeta_{\mathfrak{t}_1} &= \eta_{\mathfrak{t}_1 -}^{\nu_0} + \xi(\nu_0, \nu_{\mathfrak{t}_1}), \\
\zeta_t &= \zeta_{\mathfrak{t}_1} + \eta_{t-\mathfrak{t}_1}^{\nu_0}, \qquad \text{if} \quad t \in [\mathfrak{t}_1, \mathfrak{t}_2), \\
\zeta_{\mathfrak{t}_2} &= \zeta_{\mathfrak{t}_2 -} + \xi(\nu_{\mathfrak{t}_1}, \nu_{\mathfrak{t}_2}) \qquad \qquad \text{etc.} \\
\\
\zeta_t &= \zeta_{\mathfrak{t}_k} + \eta_{t-\mathfrak{t}_k}^{\nu_{\mathfrak{t}_k}} \qquad \text{if} \quad t \in [\mathfrak{t}_k, \mathfrak{t}_{k+1}), \\
\zeta_{\mathfrak{t}_{k+1}} &= \zeta_{\mathfrak{t}_k -} + \xi(\nu_{\mathfrak{t}_k}, \nu_{\mathfrak{t}_{k+1}}).
\end{aligned}$$

Because $\mathbf{P}_i(t_k \to \infty) = 1$, these equations define the variable, ζ_t, for any $t \geq 0$. The two dimensional process (ν_t, ζ_t) $(t \geq 0)$ is a process with independent increments in a random medium, and equation (76) is valid for $\mathbf{P}_{t,0} = \mathbf{P}_i$. The required properties of the trajectories of the process (ν_t, ζ_t) $(t \geq 0)$ follow from the corresponding properties of the processes ν_t $(t \geq 0)$ and η_t^i $(t \geq 0,\ i = 1, \ldots, d)$.

47. In what follows it will be essential that the following additional assumptions are valid.

All states of the process ν_t $(t \geq 0)$ are connectable and;

$$\text{if} \quad a_i^2 = 0, \quad \text{then} \quad \int_0^1 y l_i(dy) = \infty, \qquad i = 1, \ldots, d \tag{84}$$

These assumptions are considered to be true until the end of the chapter.

Condition (84) implies that the variation of every process η_t^i $(t \geq 0,\ i = 1, \ldots, d)$ from (83) is locally unbounded.

These assumptions simplify a number of technicalities in the subsequent consideration, but do not change anything in principle.

Distributions of the main limit functionals connected with the departure of the trajectories of the process ζ_t $(t \geq 0)$ from a finite or infinite interval are determined. All these functionals are expressed in terms of the *resolvent*, which is a matrix, $R_s(x)$ $(s \geq 0)$, continuously dependent on x $(x > 0)$, defined by its Laplace transform:

$$\int_0^\infty e^{px} R_s(x) dx = \{A(p) + sE\}^{-1}, \qquad s \geq 0, \tag{85}$$

where the absolute value of p $(p < 0)$ is sufficiently large.

48. It is not trivial to demonstrate that this definition is correct and a number additional notions are required for this. First of all, a class $\mathbb{K}$, is introduced, which consists of $d \times d$ matrices, $K(p) = \|k_{ij}(p)\|_{i,j=1}^d$, dependent on $p < 0$ such that;

$$k_{ij}(p) = \int_0^\infty e^{px} m_{ij}(dx) \qquad \text{for} \quad i \neq j \quad \text{and}$$
$$k_{ij}(p) = c_i^0 + c_i^1 p + \int_0^\infty (e^{px} - 1) m_i(dx),$$

where the measures $m_{ij}(dx)$ $(i \neq j)$ are finite, $c_i^1 \geq 0$ and $\int_0^\infty (x \wedge 1) m_i dx < \infty$.

A matrix, $K(p)$, belongs to $\mathbb{K}$ if and only if, for all $t > 0$, all elements of the matrix $\exp\{tK(p)\}$ are Laplace transforms of finite measures on $[0, \infty)$.

The construction of the resolvent requires the matrix $Q(p) \in \mathbb{K}$, which is a unique solution, in the class $\mathbb{K}$, of the equation;

$$A(Q) + pE = 0, \tag{86}$$

where $A(Q) = A_0 + \int_0^\infty L_0(dy)e^{yQ} + A_1Q + A_2Q^2 + \int_0^\infty L_1(dy)[e^{yQ} - E - (y \wedge 1)Q]$ (this expression is defined if the spectrum of the matrix Q belongs to the left half of the complex plane).

A solution of equation (86), in the class $\mathbb{K}$, can be constructed as follows. Let:

$$\theta = \inf\{t : \zeta_t \le 0\}, \quad \theta(y) = \inf\{t : \zeta_t \le y\},$$
$$G(x) = \|\mathbf{P}_{i,x}(e^{s\theta}, \nu_\theta = j)\|_{i,j=1}^d, \qquad x > 0.$$

Then $\theta(y) \le \theta$ for $y > 0$, $\theta - \theta(y) = T_{\theta(y)}\theta$ and $\nu_\theta = T_{\theta(y)}\nu_\theta$, where T_t is the operator of translation, by t, of the trajectory of the process (ν_t, ζ_t) $(t \ge 0)$. Hence;

$$\begin{aligned}\mathbf{P}_{i,x+y}(e^{s\theta}, \nu_\theta = j) &= \mathbf{P}_{i,x+y}[e^{s\theta(y)}T_{\theta(y)}(e^{s\theta}, \nu_\theta = j)] \\ &= \mathbf{P}_{i,x+y}[e^{s\theta(y)}\mathbf{P}_{\nu_{\theta(y)},y}(e^{s\theta}, \nu_\theta = j)] \\ &= \sum_k \mathbf{P}_{i,x+y}[s^{\theta(y)}, \nu_{\theta(y)} = k]\mathbf{P}_{k,y}(e^{s\theta}, \nu_\theta = j).\end{aligned}$$

However, $\mathbf{P}_{i,x+y}[e^{s\theta(y)}, \nu_{\tau(y)} = k] = \mathbf{P}_{i,x}[e^{s\tau}, \nu_\tau = k]$ and, consequently;

$$G(x+y) = G(x)G(y).$$

It follows from condition (84) that $G(0+) = E$. Hence;

$$G(x) = \exp\{xQ(s)\}. \tag{87}$$

Let $\mathfrak{A}$ be a *characteristic operator* of the process (ν_t, ζ_t) $(t \ge 0)$. The operator $\mathfrak{A}$ is defined on some class of functions of two variables, $f(i,x)$ $(i = 1, \dots, d)$, and $x \in (-\infty, \infty)$. It is convenient to consider the operator $\mathfrak{A}$ as acting on a vector $\mathbf{f}(x) = (f(i,x))_{i=1}^d$ with $\mathfrak{A}\mathbf{f}(x) = (\mathfrak{A}f(i,x))_{i=1}^d$. It is convenient to define the action of the operator, $\mathfrak{A}$, on the matrices that have d rows and are dependent on x. With this notation;

$$\mathfrak{A}G(x) + sG(x) = 0, \qquad x > 0. \tag{88}$$

49. Lemma. *If a function $\mathbf{f}(x) = (f(i,x))_{i=1}^d$ $(x \in \mathbf{R}^1)$ is twice continuously differentiable for $x > 0$ and $\lim_{x\to\infty} \mathbf{f}(x) = \lim_{x\to\infty} \mathbf{f}'(x) = \lim_{x\to\infty} \mathbf{f}''(x) = 0$, then;*

$$\begin{aligned}\mathfrak{A}\mathbf{f}(x) &= A_0\mathbf{f}(x) + \int_0^\infty L_0(dy)\mathbf{f}(x+y) + A_1\mathbf{f}'(x) + A_2\mathbf{f}''(x) \\ &\quad + \int_0^\infty L_1(dy)[\mathbf{f}(x+y) - \mathbf{f}(x) - (y \wedge 1)\mathbf{f}'(x)]\end{aligned} \tag{89}$$

for $x > 0$.

Proof. Let $\mathfrak{B}$ be an *infinitesimal operator* of the process, (ν_t, ζ_t) $(t \ge 0)$. By definition, a function, $g(j,x)$ $(j = 1, \dots, d,\ x \in \mathbf{R}^1)$, belongs to the domain of the operator, $\mathfrak{B}$, if and only if the limit:

$$\lim_{t\downarrow 0} \frac{1}{t}[\mathbf{P}_{j,x}g(\nu_t, \zeta_t) - g(j,x)] = \mathfrak{B}g(j,x)$$

exists and is uniform in $x \in \mathbf{R}^1$.

Let $g(j,x)$ be twice differentiable in x and $|g(j,x)| + |g'(j,x)| + |g''(j,x)| \to 0$ as $|x| \to \infty$. Then there exists a sequence, $g_n(j,x)$ ($n = 1,2,\ldots$), of smooth, i.e. infinitely differentiable in x, functions with compact supports such that:

$$\sup_{j,x}\{|g(j,x) - g_n(j,x)| + |g'(j,x) - g'_n(j,x)| \\ +|g''(j,x) - g''_n(j,x)|\} \underset{n\to\infty}{\to} 0.$$

Because the function g_n is infinitely differentiable and is equal to zero outside an interval of finite length, its Fourier transform;

$$\varphi_n(j,y) = \int_{-\infty}^{\infty} e^{\iota yx} g_n(j,x)dx \qquad (\iota^2 = -1),$$

is absolutely integrable with the weight, y^k, for any $k = 0,1,2,\ldots$. Indeed;

$$y^k\varphi_n(j,y) = \int_{-\infty}^{\infty} e^{\iota yx}\frac{d^k}{dx^k}g_n(j,x)dx$$

and:

$$\|y^k\varphi_n(j,y)| \leq \int_{-\infty}^{\infty}\left|\frac{d^k}{dx^k}g_n(j,x)\right|dx.$$

Consequently;

$$g_n(j,x) = \frac{1}{2\pi}\int_{-\infty}^{\infty} e^{-\iota xy}\varphi_n(j,y)dy$$

and:

$$(\mathbf{P}_{j,x}g_n(\nu_i,\zeta_t))_{j=1}^d = \frac{1}{2\pi}\int_{-\infty}^{\infty} e^{-\iota xy}e^{tA(-\iota y)}\varphi_n(y)dy, \tag{90}$$

where $\varphi_n(y) = (\varphi_n(j,y))_{j=1}^d$.

Elements of the matrix $A(-\iota y)$ do not diverge faster than y^2 as $|y|$ tends to ∞. Hence, the expression;

$$\frac{1}{2\pi}\int_{-\infty}^{\infty} e^{\iota xy}\left\{\frac{1}{t}[e^{tA(-\iota y)} - E] - A(-\iota y)\right\}\varphi_n(y)dy$$

approaches zero uniformly as $y \downarrow 0$, and (90) demonstrates that the function $\mathbf{g}_n$ belongs to the domain of definition of the operator $\mathfrak{B}$ and:

$$\begin{aligned}\mathfrak{B}\mathbf{g}_n(x) &= \frac{1}{2\pi}\int_{-\infty}^{\infty} e^{-ixy}A(-iy)\varphi_n(y)dy \\ &= A_0\mathbf{g}_n(x) + \int_0^{\infty} L_0(dy)\mathbf{g}_n(x+y) + A_1\mathbf{g}'_n(x) + A_2\mathbf{g}''_n(x) \\ &\quad + \int_0^{\infty} L_1(dy)[\mathbf{g}_n(x+y) - \mathbf{g}_n(x) - (y\wedge 1)\mathbf{g}'_n(x)],\end{aligned}$$

where $\mathbf{g}_n(x) = (g_n(j,x))_{j=1}^d$ and $\mathfrak{B}\mathbf{g}_n(x) = (\mathfrak{B}g_n(j,x))_{j=1}^d$.

The conditions imposed on $g_n(j,x)$ imply that the right hand side converges uniformly in x to:

$$A_0\mathbf{g}(x) + \int_{-\infty}^{\infty} L_0(dy)\mathbf{g}(x+y) + A_1\mathbf{g}'(x) + A_2\mathbf{g}''(x) + \int_0^{\infty} L_1(dy)[\mathbf{g}(x+y) - \mathbf{g}(x) - (y \wedge 1)\mathbf{g}'(x)]. \tag{91}$$

However, the infinitesimal operator, $\mathfrak{B}$, is closed, i.e. the uniform convergence of $\mathbf{g}_n$ to $\mathbf{g}$ and $\mathfrak{B}\mathbf{g}_n$ to h implies that $\mathfrak{B}\mathbf{g} = h$.

In this case the function, g, belongs to the domain of definition of the operator $\mathfrak{B}$ and that the vector $\mathfrak{B}\mathbf{g}$ is equal to vector (91). Because the characteristic operator $\mathfrak{A}$ is an extension of the infinitesimal operator $\mathfrak{B}$, formula (89) is thus proved in the case where $f(j,x)$ ($j = 1,\ldots,d$, $x \in \mathbf{R}^1$) is twice continuously differentiable for all x and approaches zero, together with its two first order derivatives, as $|x|$ tends to ∞.

Because the process ζ_t ($t \geq 0$) does not have any negative jumps, the characteristic operator $\mathfrak{A}$ has the following property, which is called *semilocality*, and is described below.

If $\mathbf{f}(y) = \mathbf{g}(y)$ for $y > c$, then $\mathfrak{A}\mathbf{f}(x) = \mathfrak{A}\mathbf{g}(x)$ for $x > c$ as long as one side of this equation is defined.

In other words, as far as the values of $\mathfrak{A}\mathbf{f}(x)$ are concerned for $x > c$, the values of $f(y)$, for $y \leq c$, can be neglected.

If a function $\mathbf{f}(y) = (f(i,y))_{i=1}^d$ is twice continuously differentiable for $y > 0$ and vanishes as $y \to \infty$, together with its two first order derivatives, then, for any $c > 0$, it is possible to construct an even twice-continuously differentiable function, $\mathbf{g}$, which coincides with $\mathbf{h}$ on (c,∞). For such a function, $\mathbf{g}$, the value of $\mathfrak{A}\mathbf{g}$ is given by expression (91), which, for $x > c$, is equal to the right hand side of (89) and to $\mathfrak{A}\mathbf{f}(x)$ because the operator $\mathfrak{A}$ is semilocal. Because the positive number c is arbitrary, then equation (89) is valid for all $x > 0$.

Formulae (87) and (89) demonstrate that:

$$\mathfrak{A}G(x) = A(Q(s))e^{xQ(s)} \qquad \text{for} \quad x > 0,$$

which, together with (88), yields:

$$A(Q(s)) + sE = 0. \tag{92}$$

Thus the required solution of equation (88) is constructed.

50. For every fixed s ($s < 0$), the matrix;

$$K_s(p) = [A(p) + sE]\{pE - Q(s)\}^{-} \tag{93}$$

is an element of the class $\mathbb{K}$. Indeed, from the following identities:

$$\begin{aligned}\frac{A(p)-A(q)}{p-q} &= \frac{A(q)-A(0)}{q} + A_2p + \int_0^\infty dy(e^{py}-1)\int_y^\infty L(dx)e^{q(x-y)},\\ \frac{A(p)-A(q)}{p-q} &= \frac{A(p)-A(0)}{p} + A_2q + \int_0^\infty dy e^{py}\int_y^\infty L(dx)[e^{q(x-y)}-1],\\ &\qquad p<0, \quad q<0,\end{aligned}$$

(here and in what follows, $L(dx) = L_0(dx) + L_1(dx)$) the following can be obtained:

$$\begin{aligned}K_s(p) &= K_s(0) + A_2p + \int_0^\infty dy(e^{py}-1)\int_y^\infty L(dx)e^{(x-y)Q(s)}, \qquad p\leq 0,\\ K_s(p) &= \frac{1}{p}[A(p)-A(0)] + A_2Q(s)\\ &\quad + \int_0^\infty dy e^{py}\int_y^\infty L(dx)[e^{(x-y)Q(x)} - E], \qquad p<0. \end{aligned} \tag{94}$$

This demonstrates that the off-diagonal elements of the matrix $K_s(p)$ are equal to the off-diagonal elements of the matrix:

$$\begin{aligned}&\int_0^\infty e^{py}L_0[y,\infty)dy + A_2Q(s) + \int_0^\infty e^{py}dy\int_y^\infty L(dx)[e^{(x-y)Q(s)} - E]\\ &\qquad = A_2Q(s) + \int_0^\infty e^{py}dy\int_y^\infty L_1(dy)[e^{(x-y)Q(s)} - E]\\ &\qquad\quad + \int_0^\infty e^{py}dy\int_y^\infty L_0(dy)e^{(x-y)Q(s)}.\end{aligned}$$

However, the off-diagonal elements of the latter matrix are equal to the off-diagonal elements of the matrix:

$$A_2Q(s) + \int_0^\infty e^{py}dy\int_y^\infty L(dx)e^{(x-y)Q(s)}.$$

Because the off-diagonal elements of the matrix $Q(s)$ are nonnegative, it follows from (94) that $K_s(p) \in \mathbb{K}$.

The spectrum of the matrix, $A(0)$, lies in the left half of the complex plane and $A(0)\mathbf{1} = \mathbf{0}$. Let ρ be the left annulling vector for the matrix $A(0)$, i.e. $A(0)^\top\rho = 0$ and $\langle\rho, 1\rangle = 1$. Because all states of the process, ν_t $(t \geq 0)$, are connecting, the matrix $A(0)$ is indecomposable, so that the vector ρ has positive coordinates, is unique, and coincides with the vector of the stationary probabilities of the process ν_t $(t \geq 0)$.

For $s < 0$, (93) gives the inequality $K_s(0)^\top\rho \leq -s\{Q(s)^{-1}\}^\top\rho$. If $s < 0$, the matrix $Q(s)^{-1}$ has negative elements in accordance with (84) and (87). Consequently, $K_s(0)^\top\rho < 0$ for $s < 0$. This demonstrates that the Perron's root of the matrix $K_s(0)$ is negative for $s < 0$. Hence the equation;

$$\int_0^\infty e^{py}H_s(dy) = -\{K_s(p)\}^{-1}, \qquad p<0 \quad s\leq 0,$$

defines a matrix, $H_s(dy)$, the elements of which are measures on $[0,\infty)$, which are finite for $s<0$.,

It can be proved that solutions of equation (92) are unique in the class $\mathbb{K}$. Indeed, if $Q^*(s)$ is a solution of equation (92) from the class $\mathbb{K}$ which, generally speaking, is different from $Q(s)$, then there is a corresponding matrix $K_s^*(P)\in\mathbb{K}$ and, if $\Re p=0$ and $s<0$, then:

$$A(p)+sE=K_s(p)\{pE-Q(s)\}=K_s^*(p)\{pE-Q^*(s)\}.$$

Further;

$$K_s(p)^{-1}K_s^*(p)=\{pE-Q(s)\}\{pE-Q^*(s)\}^{-1},\qquad \Re p=0,\quad s<0. \tag{95}$$

The left hand side can be analytically extended into the left half of the complex plane, $\{p:\Re p<0\}$, and the right hand side into the right half of the complex plane, $\{p:\Re p>0\}$. Thus both sides of equation (95) are parts of a single function, which is analytic in the complex plane and does not diverge faster than $|p|$ as $|p|$ tends to ∞. By Liouville's theorem, this function is thus a linear polynomial. Consequently, there exist matrices, $B(s)$ and $C(s)$, such that $\{pE-Q(s)\}\{pE-Q^*(s)\}^{-1}=pB(s)+C(s)$ $(\Re p\geq 0)$, or that;

$$pE-Q(s)=\{pB(s)+C(s)\}\{pE-Q^*(s)\}.$$

It follows that $B(s)=0$, $C(s)=E$, and $Q(s)=Q^*(s)$.

Let p $(p<0)$ be such that the half-plane $\{z:\Re z<p\}$ does not contain the spectrum of the matrix $Q(s)$.

Then $\{pE-Q(s)\}^{-1}=\int_0^\infty e^{py}e^{-yQ(s)}dy$ and, for such p, equation (85) is valid with:

$$R_s(x)=\int_0^x e^{(y-x)Q(s)}H_s(dy). \tag{96}$$

51. Thus the resolvent, $R_s(x)$, has been constructed and is continuous in x. Moreover, at every point $x>0$, the resolvent $R_s(x)$ has one-sided derivatives. The left derivative is left continuous, the right derivative is right continuous. Indeed, for the matrix $K_s(p)=\|k_{ij}^s(p)\|_{ij=1}^d=[A(p)+sE]\{pE-Q(s)\}^{-1}$, there exist nondecreasing left continuous functions, $\varphi_i^s(x)$ $(x\geq 0)$, and measures, $m_{ij}^s(dy)$ on $[0,\infty)$, such that:

$$\begin{gathered}\varphi_i^s(0)=0,\qquad \int_0^\infty e^{px}d\varphi_i^s(x)=-\frac{1}{k_{ii}^s(p)},\\ \int_0^\infty e^{py}m_{ij}^s(dy)=-\frac{k_{ij}^s(p)}{k_{ii}^s(p)},\qquad i\neq j.\end{gathered} \tag{97}$$

Laplace transforms are the easiest way to verify the following equation:

$$H_s(x)=\Phi_s(x)+\int_0^\infty M_s(dy)H_s(x-y), \tag{98}$$

where $H_s(x) = H_s[0,x)$, $\Phi_s(x) = \|\delta_{ij}\varphi_i^s(x)\|_{i,j=1}^d$, $M_s(dy) = \|(1-\delta_{ij})m_{ij}^s(dy)\|_{i,j=1}^d$. The one-sided differentiability of the resolvent $R_s(x)$ and the one-sided continuity of its derivatives follow from the absolute continuity of the functions $\varphi_i(x)$ for $x > 0$ and from the existence of one-sided limits of the derivatives at every point $x > 0$.

If $a_i^2 > 0$, then $\varphi_i^s(x)$ is continuously differentiable for $x > 0$. Indeed, let $\pi_i^s(y)$ $(i = 1, \dots, d)$ be the diagonal elements of the matrix $\int_y^\infty L(dy)e^{(x-y)Q(s)}$. Then the function;

$$\psi_i^s(x) = \frac{1}{a_i^2}\left\{1 + k_{ii}^s(0)\varphi_i^s(x) + \int_0^\infty [\varphi_i^s(x+y) - \varphi_i(x)]\pi_i^s(y)dy\right\},$$

is continuous for $x > 0$ and $\int_0^x \psi_i^s(y)dy = \varphi_i^s(x)$, as can be verified by using the Laplace transform.

If $a_i^2 = 0$, then:

$$\varphi_i^s(x) - \varphi_i^s(\epsilon) = \int_\epsilon^x dy \int_0^{y-\epsilon} d\varphi_i^s(z) \int_0^\epsilon d\varphi_i^s(u)\pi_i^s(y-z-u) \tag{99}$$

for $x \geq \epsilon > 0$. This can be verified by using a monotone process, ξ_t $(t \geq 0)$, with independent increments such that:

$$\xi_0 = 0, \qquad \mathbf{P}e^{p\xi_t} = \exp\{t[k_{ii}(p) - k_{ii}^s(0)]\}.$$

If $\tau_u = \inf\{t : \xi_t \geq u\}$, then:

$$\varphi(u) = \mathbf{P}\int_0^{\tau_u} e^{kt}dt, \qquad k = k_{ii}^s(0), \qquad \varphi(u) = \varphi_i^s(u)$$

(the dependence of i on s is not important at this stage).

By the formula of complete probability;

$$\varphi(x) = \varphi(x \wedge \epsilon) + \int_0^x n_\epsilon(dy)\varphi(x-y),$$

where $\epsilon > 0$, $n_\epsilon(dy) = \mathbf{P}(e^{k\tau_\epsilon}, \xi_{\tau_\epsilon} \in dy)$.

To obtain (99), it is sufficient to notice that $a_i^2 = 0$ implies $\mathbf{P}(\xi_{\tau_\epsilon} = \epsilon) = 0$ and then apply the formula:

$$\mathbf{P}(e^{k\tau_\epsilon}, \xi_{\tau_\epsilon} \geq y) = \int_0^\epsilon d\varphi(x)\int_{y-x}^\infty \pi(u)du, \qquad y > \epsilon. \tag{100}$$

52. Besides (96), the resolvent, $R_s(x)$, has another representation. In some sense it is a dual to (96) and is connected with a root, $Q^\bullet(s) \in \mathbb{K}$, of the equation;

$$sE + A_0 + \int_0^\infty e^{yQ^\bullet} L_0(dy) + Q^\bullet A_1 + Q^{\bullet 2}$$

$$+\int_0^\infty [e^{yQ^\bullet} - E - (y \wedge 1)Q^\bullet]L_1(dy) = 0. \tag{101}$$

If a solution to this equation belonging to the class $\mathbb{K}$ is demonstrated to exist, then by repeating the previous argument, this solution will be unique and:

$$R_s(x) = \int_0^x H_s(dy)e^{yQ^\bullet(s)},$$

where;

$$\begin{aligned} \int_0^\infty e^{py}H_s^\bullet(dy) &= -\{K_s^\bullet(p)\}^{-1}, \qquad p<0, & (102)\\ K_s^\bullet(p) &= \{pE - Q^\bullet(s)\}^{-1}[A(p) + sE] \\ &= K_s^\bullet(0) + A_2p + \int_0^\infty (E^{py} - 1)dy\int_y^\infty e^{(x-y)Q^\bullet(s)}L(dy). & (103)\end{aligned}$$

A solution to equation (101), $Q^\bullet(s)$, from the class $\mathbb{K}$, can be constructed as follows. Let $(\nu_t^\#, \zeta_t^\#)$ $(t \geq 0)$ be a process with independent increments in a random medium such that:

$$\|\mathbf{P}_{i,0}(e^{s\zeta_t^\#}, \nu_t^\# = j)\|_{i,j=1}^d = \exp\{tA^\#(s)\},$$

where $A^\#(s) = P^{-1}A(s)^\top P$ and the elements in the diagonal of the matrix P are equal to the coordinates of the vector ρ that annuls the matrix $A(0)$. It is clear that the process $\zeta_t^\#$ $(t \geq 0)$ has only positive jumps and that all of the processes $\nu_t^\#$ $(t \geq 0)$ are connecting. Moreover, the diagonal elements of the matrices $A^\#(s)$ and $A(s)$ coincide. Let the matrix $Q^\#(s)$ be constructed from the process $(\nu_t^\#, \zeta_t^\#)$ $(t \geq 0)$ in the same way as the matrix $Q(s)$ is constructed from the process (ν_t, ζ_t) $(t \geq 0)$. Then, the solution of equation (101) is $Q^\bullet(s) = P^{-1}Q^\#(s)^\top P$.

53. Calculation of the distributions of the limit functionals is based on the formula:

$$\begin{aligned} \left(\mathbf{P}_{i,x}\int_0^{\theta_z} e^{st}f(\nu_t,\zeta_t)dt\right)_{i=1}^d &= R_t(z-x)\int_{-\infty}^z e^{(z-y)Q^\bullet(s)}\mathbf{f}(y)dy \\ &\quad - \int_x^z R_s(y-x)\mathbf{f}(y)dy, & (104)\end{aligned}$$

where $s<0$, $\theta_z = \inf\{t : \zeta_t \geq z\}$ and $\mathbf{f}(y) = (f(i,y))_{i=1}^d$ $(y<z)$ is a bounded Borel function.

It is sufficient to prove formula (104) in a particular case, where $z = 0$, $f(i,y) = e^y p\delta_{ij}$ and $\Re p = 0$. In this case, the formula takes the form:

$$\left\|\mathbf{P}_{i,-x}\int_0^{\theta_0} e^{st}e^{p\zeta_t}i_{\{\nu_t=j\}}dt\right\|_{i,j=1}^d = R_s(x)\{pE - Q^\bullet(s)\}^{-1}$$

$$- \int_0^x R_s(x-y)e^{-py}dy, s<0,\ x>0. \quad (105)$$

The left hand side of this equation, $V(x,p)$, is bounded and at least right continuous for $x > 0$. Hence (105) follows from:

$$\int_0^\infty e^{qx}V(x,p)dx$$
$$= \{A(q)+sE\}^{-1}\left[\{pe - Q^\bullet(s)\}^{-1} - \frac{1}{p-q}E\right], \quad q<0. \qquad (106)$$

Applying (76);

$$\left\| \mathbf{P}_{i,-x}\int_0^\infty e^{st}e^{p\zeta_t}i_{\{\nu_t=j\}}dt \right\|_{i,j=1}^d = \int_0^\infty e^{st}e^{-xp}e^{tA(p)}$$
$$= -e^{xp}\{A(p)+sE\}^{-1}.$$

Further;

$$\mathbf{P}_{i,-x}\int_0^\infty e^{st}e^{p\zeta_t}i_{\{\nu_t=j\}}dt = \mathbf{P}_{i,-x}\int_0^{\theta_0} e^{st}e^{p\zeta_t}i_{\{\nu_t=j\}}dt$$
$$+ \mathbf{P}_{i,-x}\left[e^{s\theta_0}\mathbf{P}_{\nu_{\theta_0},\zeta_{\theta_0}}\int_0^\infty e^{st}e^{p\zeta_t}i_{\{\nu_t=j\}}dt\right]$$

and, consequently;

$$-e^{-xp}\{A(p+sE\}^{-1} = V(x,p) - U(x,p)\{A(p)+sE\}^{-1},$$

where $U(x,p) = \|\mathbf{P}_{i,-x}(e^{s\theta_0}e^{p\zeta_{\theta_0}}, \nu_{\theta_0}=j)\|_{i,j=1}^d$.

By taking the Laplace transform, the following is obtained:

$$\frac{1}{q-p}\{A(p)+sE\}^{-1} = V^*(q,p) - U^*(q,p)\{A(p)+sE\}^{-1},$$

where $V^*(q,p) = \int_0^\infty e^{qy}V(y,p)dy$, $U^*(q,p) = \int_0^\infty e^{qy}U(y,p)dy$.

This, together with (103), yields:

$$\frac{1}{q-p}[\{K_s(q)\}^{-1} - \{K_s(p)\}^{-1}] - U^*(q,p)\{K(p)\}^{-1}$$
$$\frac{1}{q-p}\{K_s(q)\}^{-1} - V^*(q,p)\{pE - Q^\bullet(s)\}. \qquad (107)$$

From (102) the matrices $U^*(q,p)$ and $\{K_s(p)\}^{-1}$, considered as functions of p, allow the analytic continuation into the left half of the complex plane $\{p : \Re p < 0\}$, and the matrix $V^*(q,p)$ allows the continuation into the right half $\{p : \Re p > 0\}$.

This implies that the left hand side of (107) is analytic and bounded on the left of the imaginary axis, $\{p : \Re p = 0\}$, and is bounded on the right. By Liouville's theorem, this is only possible if both sides of (107) do not depend on p, i.e. there exists a matrix, $C(q)$, such that:

$$\frac{1}{q-p}\{K_s(q)\}^{-1} - V^*(q,p)\{pE - Q^\bullet(s)\} = C(q), \qquad \Re p \geq 0, \tag{108}$$

and:

$$\frac{1}{q-p}[\{K_s^\bullet(q)\}^{-1} - \{K_s^\bullet(p)\}^{-1}] - U^*(q,p)\{K_s^\bullet(p)\}^{-1} = C(q), \qquad \Re p \leq 0. \tag{109}$$

However;

$$\lim_{p\to-\infty}\{K_s^\bullet(p)\}^{-1} = -\lim_{p\to\infty} p\{A(p)+sE\}^{-1} = 0$$

because the diagonal elements of the matrix $\frac{1}{p}A(p)$, according to (84), approach $-\infty$ as p tends to $-\infty$.

By taking the limit, $p \to \infty$ in (109), then $C(q) = 0$ and (108) yields the following equation, which is equivalent to (106);

$$\begin{aligned} V^*(q,p) &= \frac{1}{q-p}\{K_s^\bullet(q))\}^{-1}\{pE - Q^\bullet(s)\}^{-1} \\ &= \frac{1}{q-p}\{A(q)+sE\}^{-1}\{qE - Q^\bullet(s)\}\{pE - Q^\bullet(s)\}^{-1}. \end{aligned}$$

54. Formula (104) gives the following two formulae, which are connected with the time $\theta_z^0 = \inf\{t : \zeta_t \notin (0,z)\}$, $z > 0$:

$$\begin{aligned} &\|\mathbf{P}_{i,x}(e^{s\theta_z^0}, \zeta_{\theta_z^0} = 0, \nu_{\theta_z^0} = j)\|_{i,j=1}^d \\ &\quad = R_s(z-x)R_s(z)^{-1}, \qquad x \in (0,z), \quad s \leq 0; \end{aligned} \tag{110}$$

$$\begin{aligned} &\left(\mathbf{P}_{i,x}\int_0^{\theta_z^0} e^{st} f(\nu_t,\zeta_t)dt\right)_{i=1}^d \\ &\quad R_s(z-x)R_s(z)^{-1}\int_0^z R_s(y)\mathbf{f}(y)dy - \int_x^z R_s(y-x)\mathbf{f}(y)dy, \end{aligned} \tag{111}$$

where $x \in (0,z)$, $s \leq 0$ and the function $\mathbf{f} = (f(i,y))_{i=1}^d$ is measurable and bounded.

If $f(i,y) = i_{(-\infty,0]}(y)h(i)$ is substituted in (104), then for $x \in (0,z)$, the right hand side becomes $-R_s(z-x)Q^\bullet(s)^{-1}e^{zQ^\bullet(s)}\mathbf{h}$, where $\mathbf{h} = (h(i))_{i=1}^d$. Further;

$$\begin{aligned} \mathbf{P}_{i,x}\int_0^{\theta_z} e^{st}f(\nu_t,\zeta_t)dt &= \mathbf{P}_{i,x}\int_0^{\theta_z^0} e^{st}f(\nu_t,\zeta_t)dt \\ &+ \mathbf{P}_{i,x}e^{s\theta_z^0}\mathbf{P}_{\nu_{\theta_z^0}\zeta_{\theta_z^0}}\int_0^{\theta_z} e^{st}f(\nu_t,\zeta_t)dt \\ &= \mathbf{P}_{i,x}\int_0^{\theta_z^0} e^{st}f(\nu_t,\zeta_t)dt \end{aligned}$$

$$+ \sum_j \mathbf{P}_{i,x}(e^{s\theta_x^0}, \zeta_{\theta_x^0} = 0, \nu_{\theta_x^0} = j)\mathbf{P}_{j,0} \int_0^{\theta_x} e^{st} f(\nu_t, \zeta_t) dt \quad (112)$$

For this choice of $f(i, y)$, the first term in the right hand side of the latter equation becomes zero and, consequently, if $U(x, z)$ denotes the left hand side of (110), then;

$$R_s(z - x)Q^\bullet(s)^{-1}e^{sQ^\bullet(s)}\mathbf{h} = U(x, z)R_s(z)Q^\bullet(s)^{-1}e^{sQ^\bullet(s)}\mathbf{h}.$$

Because the vector $\mathbf{h}$ is arbitrary;

$$R_s(z - x) = U(x, z)R_s^\bullet(z), \quad (113)$$

and (110) is reduced to the problem of whether the matrix, $R_s(z)$, can be inverted. It follows from (113) that to demonstrate that $R_s(z)$ is invertible then it is only sufficient to prove the fact for sufficiently small $x > 0$. Equation (98), together with relations (94) and (97), demonstrates that:

$$\frac{m_{ij}^s[0, x)}{\varphi_i^s(x)} \to 0 \quad \text{as} \quad x \downarrow 0.$$

Hence, $\Phi_s(x)^{-1}H_s(x) \to E$ as $x \downarrow 0$ and, from (96), $\Phi_s(x)^{-1}R_s(x) \to E$ as $x \downarrow 0$.

The matrix $R_s(z)$ is thus invertible for sufficiently small (and hence for all) $x > 0$. This demonstrates the validity of formula (110).

If the right hand sides of (104) and (110) are substituted into (112), then the left hand side of (111) is thus equal to:

$$R_z(z - x) \int_{-\infty}^{z} e^{(z-x)Q^\bullet(s)}\mathbf{f}(y)dy - \int_x^z R_s(y - x)\mathbf{f}(y)dy$$
$$-R_s(z - x)R_s(z)^{-1} \left\{ R_s(z) \int_{-\infty}^{z} e^{(z-x)Q^\bullet(s)}\mathbf{f}(y)dy - \int_0^z R_s(y)\mathbf{f}(y)dy \right\},$$

which is identical to the right hand side of (111).

3.2 Boundary-value conditions

55. The problem of describing random walks on a semiaxis, such walks being related to a given process with independent increments in a random medium, (ν_t, ζ_t) $(t \geq 0)$, can be formulated as follows.

To describe all ways in which the process (ν_t, ζ_t) $(t < \infty)$, where $\theta = \inf\{t : \zeta_t \leq 0\}$, can be extended to a (homogeneous) strictly Markovian Feller process, $(\mathfrak{n}_t, \mathfrak{z}_t)$ $(t \geq 0)$, trajectories of which in the phase space, $\{1, \ldots, d\} \times [0, \infty)$, are continuous.

A process, $(\mathfrak{n}_t, \mathfrak{z}_t)$ $(t \geq 0)$, is called an extension of the process (ν_t, ζ_t) $(t < \theta)$, if the distribution of the latter process coincides with the distribution of the process $(\mathfrak{n}_t, \mathfrak{z}_t)$ $(t < \tau)$, where $\tau = \inf\{t : \zeta_t = 0\}$.

Suppose such extensions exist and let $(\mathfrak{n}_t, \mathfrak{z}_t)$ $(t \geq 0)$ be one of them. Then, if a bounded continuous function, $f(i,x)$ $(i = 1, \ldots, d,\ x \geq 0)$, approaches zero as x tends to ∞, then:

$$\begin{aligned} |\mathbf{P}_{i,x} f(\mathfrak{n}_t, \mathfrak{z}_t)| &\leq \mathbf{P}_{i,x}[|f(\mathfrak{n}_t, \mathfrak{z}_t)|, t < \tau] + \mathbf{P}_{i,x}[|f(\mathfrak{n}_t, \mathfrak{z}_t)|, \tau \leq t] \\ &\leq \mathbf{P}_{i,x}[|f(\nu_t, \zeta_t)|, t < \theta] + \sup_{j,y} |f(j,y)| \mathbf{P}_{i,x}(\theta \leq t) \underset{x \to \infty}{\longrightarrow} 0. \end{aligned}$$

Hence, the semi-group of operators, which correspond to the process $(\mathfrak{n}_t, \mathfrak{z}_t)$ $(t \geq 0)$, maps the space, $\mathbb{C}$, of the continuous functions $f(i,x)$ $(i = 1, \ldots, d,\ x \geq 0)$, which have a limit at infinity, onto itself. In other words, the process $(\mathfrak{n}_n, \mathfrak{z}_t)$ $(t \geq 0)$, is a regular Markovian process in a locally compact space. If $\mathfrak{G}$ is a characteristic operator of the process, $(\mathfrak{n}_t, \mathfrak{z}_t)$ $(t \geq 0)$, then the domain, $\mathbb{D}$, of the infinitesimal operator, $\mathfrak{g}$, of the process $(\mathfrak{n}_t, \mathfrak{z}_t)$ $(t \geq 0)$, consists of precisely the functions $f \in \mathbb{C}$, for which $\mathfrak{G} f \in \mathbb{C}$. In particular;

$$\mathfrak{G} f(i,0) = \lim_{x \downarrow 0} \mathfrak{G} f(i,x).$$

Let $\mathbb{C}_2$ be the set of those functions from $\mathbb{C}$, where the two first order derivatives of which vanish at infinity. Because the process $\mathfrak{z}_t$ $(t \geq 0)$, does not have negative jumps, the analytical form of the function $\mathfrak{G} f(i,x)$ $(x > 0,\ f \in \mathbb{C}_2)$ does not depend on the way it was extended and $\mathfrak{G} f(i,x) = \mathfrak{A} f(i,x)$, where $\mathfrak{A}(i,x)$ is equal to the ith coordinate of the right-hand side of (89) for $x > 0$. Hence, each extension, $(\mathfrak{n}_t, \mathfrak{z}_t)$ $(t \geq 0)$, is uniquely defined by its domain, $\mathbb{D}$. It is not necessary to describe the entire the set $\mathbb{D}$, but only a dense (in the uniform topology) subset, $\mathbb{D}^{\bullet}$, of it, which has the following property: for some $s < 0$, the equation $\mathfrak{G} g + sg = f$ has a solution, $g \in \mathbb{D}^{\bullet}$, for a dense (in $\mathbb{C}$) set of functions, f.

56. Theorem. *If a set, $\mathbb{D} \subset \mathbb{C}$, is a domain of the infinitesimal operator describing a random walk on a semiaxis, then the set, $\mathbb{D} \cap \mathbb{C}_2$, contains precisely the functions, f, from $\mathbb{C}_2$, which satisfy the boundary-value conditions;*

$$\begin{aligned} B_2 \mathfrak{A} \mathbf{f}(0) &= B_0 \mathbf{f}(0) + \int_0^\infty M_0(dy) \mathbf{f}(y) + B_1 \mathbf{f}'(0) \\ &\quad + \int_0^\infty M_1(dy)[\mathbf{f}(y) - \mathbf{f}(0)], \end{aligned} \tag{114}$$

where $\mathbf{f}(y) = (f(i,y))_{i=1}^d$; the diagonal matrices B_0, B_1, B_2 and $M_1(dy)$ have the elements b_i^0, b_i^1, b_i^2 and $m_i(dy)$ respectively; the diagonal elements of the matrix

$M_0(dy) = \|m_{ij}(dy)\|_{i,j=1}^{d}$ *are equal to zero;*

$$\sum_{j \neq i} M_{ij}[0,\infty) + b_i^0 = 0; \tag{115}$$

$$m_i\{0\} = 0, \qquad \int_0^\infty (y \wedge 1) m_i(dy) < \infty \tag{116}$$

and, if $b_i^1 = b - i^2 = 0$*, then* $m(0,\infty) = +\infty$. *(117)*

Proof. Because $\mathbb{D} = \{f \in \mathbb{C} : \mathfrak{G} f \in \mathbb{C}\}$ and $\mathfrak{G} f(i,x) = \mathfrak{A} f(i,x)$ for $x > 0$, then:

$$\mathbb{D} \cap \mathbb{C}_2 = \{f \in \mathbb{C}_2 : \mathfrak{G} f(i,0) = \mathfrak{A} f(i,0+),\ i = 1,\ldots,d\}.$$

Let the state $(i,0)$ be nonabsorbing. A bounded measurable function, $g(i,x)$ ($i = 1,\ldots,d$, $x \geq 0$), is defined to belong to the domain of the characteristic operator, $\mathfrak{G}$, at the point $(i,0)$ if the following limit exists:

$$\mathfrak{G} g(i,0) = \lim_{z \downarrow 0} \frac{\mathbf{P}_{i,0} g(\mathfrak{n}_{\tau_z \wedge \mathfrak{s}}, \mathfrak{z}_{\tau_z \wedge \mathfrak{s}}) - g(i,0)}{\mathbf{P}_{i,0} \tau_z \wedge \mathfrak{s}}, \tag{118}$$

where $\mathbf{P}_{i,0}$ is the conditional probability for $\mathfrak{n}_0 = i$, $\mathfrak{z}_0 = 0$ and:

$$\tau_z = \inf\{t : \zeta_t \geq z\}, \qquad \mathfrak{s} = \inf\{t : \mathfrak{n}_t \neq \mathfrak{n}_0\}.$$

The first term in the numerator of the fraction in the right-hand side of (118) is equal to:

$$\mathbf{P}_{i,0}[g(i,\mathfrak{z}_{\tau_z}), \tau_z < \mathfrak{s}] + \mathbf{P}_{i,0}[g(\mathfrak{n}_\mathfrak{s}, \mathfrak{z}_\mathfrak{s}), \mathfrak{s} \leq \tau_z]. \tag{119}$$

Thus the first term is equal to:

$$\mathbf{P}_{i,0}\{\tau_y < \mathfrak{s}, \mathbf{P}_{i,\mathfrak{z}_{\tau_y}}[g(i,\mathfrak{z}_{\tau_z}), \tau_z < \mathfrak{s}]).$$

Let τ_z^0 be the moment at which the process $\mathfrak{z}_t$ ($t \geq 0$), leaves the interval $(0,z)$ for the first time, i.e. $\tau_z^0 = \inf\{t : \mathfrak{z}_t \notin (0,z)\} = \tau \wedge \tau_z$. Then;

$$\begin{aligned}
\mathbf{P}_{i,x}[g(i,\mathfrak{z}_{\tau_z}), \tau_z < \mathfrak{s}] &= \mathbf{P}_{i,x}\{\tau_z^0 < \mathfrak{s}, \mathbf{P}_{i,\mathfrak{z}_{\tau_z^0}}[g(i,\mathfrak{z}_{\tau_z}), \tau_z < \mathfrak{s}]\} \\
&= \mathbf{P}_{i,x}[\tau_z^) < \mathfrak{s}, \mathfrak{z}_{\tau_z^0} \geq z, g(i,\mathfrak{z}_{\tau_z^0})] \\
&\quad + \mathbf{P}_{i,x}(\tau_z^0 < \mathfrak{s}, \mathfrak{z}_{\tau_z^0} = 0)\mathbf{P}_{i,0}[g(i,\mathfrak{z}_{\tau_z}), \tau_z < \mathfrak{s}].
\end{aligned}$$

Consequently;

$$\mathbf{P}_{i,0}[g(i,\mathfrak{z}_{\tau_z}), \tau_z < \mathfrak{s}] \frac{1}{c_i(y,z)} \mathbf{P}\{\tau_y < \mathfrak{s}, \mathbf{P}_{i,\mathfrak{z}_{\tau_y}}[g(i,\mathfrak{z}_{\tau_z^0}), \mathfrak{z}_{\tau_z^0} \geq z, \tau_z^0 < \mathfrak{s}]\}, \tag{120}$$

where;

$$c_i(y,z) = 1 - \mathbf{P}_{i,0}\{\tau_y < \mathfrak{s}, \mathbf{P}_{i,\mathfrak{z}_{\tau_y}}[\mathfrak{z}_{\tau_y^0} = 0, \tau_z^0 < \mathfrak{s}]\}, \tag{121}$$

and assumption (84) implies that $c_i(y,z)$ are positive for all $0 < y < z$.

Therefore;

$$c_i(y,z)c_i(z,1) = c_i(y,1) \tag{122}$$

for $0 < y < z < 1$. Indeed;

$$c_i(y,z) = 1 - \mathbf{P}_{i,0}[q_z(\mathfrak{z}_{\tau_y}), \tau_y < \mathfrak{z}],$$

where $q_z(x) = \mathbf{P}_{i,x}(\mathfrak{z}_{\tau_z^0} = 0, \tau_z^0 < \mathfrak{s})$ (it is not important at this point whether $q_z(x)$ should depend on i or not). Consequently;

$$\begin{aligned} c_i(y,1) &= 1 - \mathbf{P}_{i,0}[q_1(\mathfrak{z}_{\tau_z}), \tau_z < \mathfrak{s}] \\ &= 1 - \mathbf{P}_{i,0}\{\tau_y < \mathfrak{s}, \mathbf{P}_{i,\mathfrak{z}_{\tau_y}}[q_1(\mathfrak{z}_{\tau_z}), \tau_z < \mathfrak{s}]\}. \end{aligned} \tag{123}$$

However;

$$\begin{aligned} \mathbf{P}_{i,x}[q_1(\mathfrak{z}_{\tau_z}), \tau_z < \mathfrak{s}] &= \mathbf{P}_{i,x}\{\tau_z^0 < \mathfrak{s}, \mathbf{P}_{i,\mathfrak{z}_{\tau_z^0}}[q_1(\mathfrak{z}_{\tau_z}), \tau_z < \mathfrak{s}]\} \\ &= \mathbf{P}_{i,x}\{\tau_z^0 < \mathfrak{s}, \mathfrak{z}_{\tau_z^0} \geq z, q_1(\mathfrak{z}_{\tau_z^0})] \\ &\quad + \mathbf{P}_{i,x}(\tau_z^0 < \mathfrak{s}, \mathfrak{z}_{\tau_z^0} = 0)\mathbf{P}_{i,0}[q_1(\mathfrak{z}_{\tau_z}), \tau_z < \mathfrak{s}] \\ &= \mathbf{P}_{i,x}[\tau_z^) < \mathfrak{s}, q_1(\mathfrak{z}_{\tau_z^0})] - \mathbf{P}_{i,x}(\tau_z^) < \mathfrak{s}, \mathfrak{z}_{\tau_z^)} = 0)c_i(z,1) \\ &= q_1(x) - \mathbf{P}_{i,u}(\mathfrak{z}_{\tau_z^0} = 0, \tau_z^0 < \mathfrak{s})c_i(z,1). \end{aligned} \tag{124}$$

In the last equation, the fact that the function q_1 is *harmonic* was used, i.e.

$$q_1(x) = \mathbf{P}_{i,x}[q_1(\mathfrak{z}_{\tau_z^0}), \tau_z^0 < \mathfrak{s}], \qquad 0 < z < 1.$$

If (124) is substituted into (123), then:

$$\begin{aligned} c_i(z,1) &= 1 - \mathbf{P}_{i,0}[\tau_y < \mathfrak{s}, q_1(\mathfrak{z}_{\tau_y})] + \mathbf{P}_{i,0}[\tau_y < \mathfrak{s}, \mathbf{P}_{i,\mathfrak{z}_{\tau_y}}(\mathfrak{z}_{\tau_z^0} = 0, \tau_z^0 < \mathfrak{s})]c_i(z,1) \\ &= c_i(y,1) + [1 - c_i(y,z)]c_i(z,1), \end{aligned}$$

which is equivalent to (122).

By using (122), the relation (120) can be written as:

$$\begin{aligned} \frac{1}{c_i(z)}\mathbf{P}_{i,0}[g(i,\mathfrak{z}_{\tau_z}), \tau_z < \mathfrak{s}] &= \frac{1}{c_i(y)}\mathbf{P}_{i,0}\{\tau_y < \mathfrak{s}, \mathbf{P}_{i,\mathfrak{z}_{\tau_y}}[g(i,\mathfrak{z}_{\tau_z^0}), \mathfrak{z}_{\tau_z^0}), \mathfrak{z}_{\tau_z^0} \geq z, \tau_z^0 < \mathfrak{s}]\} \\ &= \int_0^\infty m_i^{(y)}(dx)\mathbf{P}_{i,x}[g(i,\mathfrak{z}_{\tau_z^0}), \mathfrak{z}_{\tau_z^0} \geq z, \tau_z^0 < \mathfrak{s}], \end{aligned} \tag{125}$$

where $0 < y < z \leq 1$, $c_i(z) = c_i(z,1)$ and $m_i^{(y)}(dx) = \frac{1}{c_i(y)}\mathbf{P}_{i,0}(\mathfrak{z}_{\tau_y} \in dx, \tau_y < \mathfrak{s})$.

If $x = 1$ in (125) and $g(i,x)$ is taken to be the indicator of the interval $[a,b] \subset [1,\infty)$, then:

$$\sup_{y \in (0,1)} m_i^{(y)}[a,b] \leq \mathbf{P}_{i,0}(\mathfrak{z}_{\tau_1} \in [a,b], \tau_1 < \mathfrak{s}). \tag{126}$$

Moreover, from (121);

$$1 \geq c_i(y,z) \geq \mathbf{P}_{i,0}\{\tau_y < \mathfrak{s}, [1 - \mathbf{P}_{i,\mathfrak{z}_{\tau_y}}(\mathfrak{z}_{\tau_y^0} = 0, \tau_z^0 < \mathfrak{s})]\}.$$

Therefore;

$$\int_0^1 m_i^{(y)}(dx)[1 - \mathbf{P}_{i,x}(\mathfrak{z}_{\tau_i^0} = 0, \tau_1^0 < \mathfrak{s})] \leq 1. \tag{127}$$

The distribution of the processes $\mathfrak{z}_t$, $t < \tau$, and ζ_t, $t < \theta$, coincide. Thus;

$$\mathbf{P}_{i,x}(\mathfrak{z}_{\tau_z^0} = 0, \tau_z^0 < \mathfrak{s})\mathbf{P}_{i,x}(\zeta_{\theta_z^0} = 0, \theta_z^0 < \mathfrak{t}), \tag{128}$$

where $\theta_z^0 = \inf\{t : \zeta_t \notin (0,z)\}$, $\mathfrak{t} = \inf\{t : \nu_t \neq \nu_0\}$.

According to the representation of the process (ν_t, ζ_t) $(t \geq 0)$, given in **46**, the right-hand side of (128) is equal to:

$$\mathbf{P}_{i,x}(e^{a_i^0 \theta_i(0,z)}, \eta^i_{\theta_i(0,z)} = 0), \tag{129}$$

where $\theta_i(0,z) = \inf_i\{t : \eta_t^i \notin (0,z)\}$ and the process η_t^i $(t \geq 0)$, with independent increments is given by (83). Consequently, to calculate the probability in the left-hand side of (128), the following scalar version of formula (110) can be used:

$$\mathbf{P}_{i,x}(\mathfrak{z}_{\tau_z^0} = 0, \tau_z^0 < \mathfrak{s}) = \frac{r_i(z-x)}{r_i(z)},$$

where $0 < x < z$;

$$\int_0^\infty e^{px} r_i(x)dx = \frac{1}{a_i(p)}, \tag{130}$$

$$p < q_i, \qquad a_i(q_i) = 0,$$

$$a_i(p) = a_i^0 + a_i^1 p + a_i^2 p^2 + \int_0^\infty l_i(dy)[e^{py} - 1 - p(y \wedge 1)].$$

Moreover, from the results of **51** it follows that, for $z > 0$, the left derivative;

$$r_i(z) = \lim_{y \downarrow 0} \frac{1}{y}[r_i(z) - r_i(z-y)]$$

exists, is positive, and is left continuous. Consequently;

$$0 < \lim_{x \downarrow 0} \frac{1}{x}[1 - \mathbf{P}_{i,x}(\mathfrak{z}_{\tau_z^0} = 0, \tau_z^0 < \mathfrak{s})] < \infty \qquad \text{for} \quad z > 0$$

and inequality (127) yields:

$$\sup_{y \in (0,1)} \int_0^1 x m_i^{(y)}(dx) < \infty.$$

This, together with (126), implies that there exists a sequence, y_n $(n = 1, 2, \ldots)$, which converges to zero. It also implies the existence of a number, b_i^1, and a measure m_i where $\int_0^\infty (x \wedge 1) m_i(dx) < \infty$ such that:

$$\frac{1}{c_i(y)} \mathbf{P}_{i,0}[f(\mathfrak{z}_{\tau_y}), \tau_y < \mathfrak{s}] \to b_i^1 f'(0) + \int_0^\infty m_i(dx) f(x) \tag{131}$$

is true if y, by taking the values from y_n $(n = 1, 2, \ldots)$, approaches zero, the bounded continuous function, $f(x)$ $(x \geq 0)$, is differentiable at zero and that $f(0) = 0$.

The second term in (119), for $0 < y < z$, is equal to:

$$\mathbf{P}_{i,0}[g(\mathfrak{n}_s, \mathfrak{z}_s), s < \tau_y] + \mathbf{P}_{i,0}\{\tau_y < s, \mathbf{P}_{i,\mathfrak{z}_{\tau_y}}[g(\mathfrak{n}_s, \mathfrak{z}_s), s \leq \tau_z]\}.$$

However;

$$\begin{aligned}\mathbf{P}_{i,x}[g(\mathfrak{n}_s, \mathfrak{z}_s), s \leq \tau_z] &= \mathbf{P}_{i,x}[g(\mathfrak{n}_s, \mathfrak{z}_s, s \leq \tau_z^0] + \mathbf{P}_{i,x}\{\tau_z^0 < s, \mathbf{P}_{i,\mathfrak{z}_{\tau_z^0}}[g(\mathfrak{n}_s, \mathfrak{z}_s), s \leq \tau_z]\} \\ &= \mathbf{P}_{i,x}[g(\mathfrak{n}_s, \mathfrak{z}_s), s \leq \tau_z^0] \\ &\quad + \mathbf{P}_{i,x}(\tau_z^0 < s, \mathfrak{z}_{\tau_z^0} = 0)\mathbf{P}_{i,0}[g(\mathfrak{n}_s, \mathfrak{z}_s), s \leq \tau_z].\end{aligned}$$

Consequently;

$$\begin{aligned}\frac{1}{c_i(z))}\mathbf{P}_{i,0}[g(\mathfrak{n}_s, \mathfrak{z}_s), s \leq \tau_z] &= \frac{1}{c_i(y)}\mathbf{P}_{i,0}[g(\mathfrak{n}_s, \mathfrak{z}_s), s \leq \tau_y] \\ &+ \frac{1}{c_i(y)}\mathbf{P}_{i,0}\{\tau_y < s, \mathbf{P}_{i,\mathfrak{z}_{\tau_y}}[g(\mathfrak{n}_s, \mathfrak{z}_s), s \leq \tau_z]\}.\end{aligned}$$

This equation defines a family of finite measures, m_{ij} $(i \neq j)$, on $[0, \infty)$ and implies that:

$$\frac{1}{c_i(y)}\mathbf{P}_{i,0}[g(\mathfrak{n}_s, \mathfrak{z}_s), s \leq \tau_y] \to \sum_{j \neq i} \int_0^\infty m_{ij}(dx)g(j, x)$$

and:

$$\frac{1}{c_i(y)}\mathbf{P}_{i,0}(s \leq \tau_y) \to -b_i^0 = \sum_{j \neq i} m_{ij}[0, \infty) \tag{132}$$

is true if y, by taking values from y_n $(n = 1, 2, \ldots)$, approaches zero and the function g is bounded and continuous. This, together with (131) gives:

$$\begin{aligned}\frac{1}{c_i(y)}[\mathbf{P}_{i,0}g(\mathfrak{n}_{\tau_y \wedge s}, \mathfrak{z}_{\tau_y \wedge s}) - g(i, 0)] &\to b_i^0 g(i, 0) + \sum_{j \neq i} \int_0^\infty m_{ij}(dy)g(i, y) + b_i^1 g'(i, 0) \\ &+ \int_0^\infty m_i(dy)[g(i, y) - g(i, 0)],\end{aligned} \tag{133}$$

where y, taking values from y_n $(n = 1, 2, \ldots)$, approaches zero and the function $g(j, y)$ $(j = 1, \ldots, d,\ y \geq 0)$ is continuous, bounded, and differentiable at the point $(i, 0)$.

Let $g(i, 0) = 0$. Then;

$$\begin{aligned}\frac{1}{c_i(z)}\mathbf{P}_{i,0}g(\mathfrak{n}_{\tau_z \wedge s}, \mathfrak{z}_{\tau_z \wedge s}) &= \frac{1}{c_i(y)}\mathbf{P}_{i,0}[g(\mathfrak{n}_s, \mathfrak{z}_s), s \leq \tau_y] \\ &+ \frac{1}{c_i(y)}\mathbf{P}_{i,0}[\tau_y < s, \mathbf{P}_{i,\mathfrak{z}_{\tau_y}}g(\mathfrak{n}_{\tau_z^0 \wedge s}, \mathfrak{z}_{\tau_z^0 \wedge s})].\end{aligned} \tag{134}$$

If, in addition, $g \in \mathbb{C}_2$, then, by Dynkin's formula;

$$\mathbf{P}_{i,x}g(\mathfrak{n}_{\tau_z^0 \wedge s}, \mathfrak{z}_{\tau_z^0 \wedge s}) = g(i, x) + \mathbf{P}_{i,x}\int_0^{\tau_z^0 \wedge s} \mathfrak{A}g(i, \mathfrak{z}_t)dt \tag{135}$$

(here we used the fact that the process $(\mathfrak{n}_t, \mathfrak{z}_t)$, $(t \geq 0)$ is an extension of the process (ν_t, ζ_t) $(t < \theta)$). After substituting (135) into (134);

$$\begin{aligned} \frac{1}{c_i(z)} \mathbf{P}_{i,0} g(\mathfrak{n}_{\tau_z \wedge \mathfrak{s}}, \mathfrak{z}_{\tau_z \wedge \mathfrak{s}}) &= \frac{1}{c_i(y)} \mathbf{P}_{i,0} g(\mathfrak{n}_{\tau_y \wedge \mathfrak{s}}, \mathfrak{z}_{\tau_y \wedge \mathfrak{s}}) \\ &+ \frac{1}{c_i(y)} \mathbf{P}_{i,0} [h_z(i, \mathfrak{z}_{\tau_y}), \tau_y < \mathfrak{s}], \end{aligned} \tag{136}$$

where $0 < y < z$, $h_z(i,x) = \mathbf{P}_{i,x} \int_0^{\tau_z^0 \wedge \mathfrak{s}} \mathfrak{A} g(i, \mathfrak{z}_t) dt$, $g \in \mathbb{C}_2$ and $g(i,0) = 0$.

The expression for $h_z(i,x)$ can be written in the form:

$$h_z(i,x) = \mathbf{P}_{i,x} \int_0^{\theta_i(0,z)} e^{a_i^0 t} \mathfrak{A} g(i, \eta_t^i) dt,$$

where $\theta_i(0,z)$ and η_t^i are the same as in (129).

This implies that, to calculate $h_z(i,x)$, the scalar version of formula (111) can be used;

$$h_z(i,x) = \frac{r_i(z-x)}{r_i(z)} \int_0^z r_i(y) \mathfrak{A} g(i,y) dy - \int_x^z r_i(y-x) \mathfrak{A} g(i,y) dy. \tag{137}$$

Here, $x \in (0,z)$ and $r_i(x)$ are defined by (130). It follows that $h_z(i,x)$ is differentiable with respect to x at zero. Moreover, $h_z(i,x)$ is continuous in x. This is clear for $x \neq z$ and, if $x = z$, it follows from:

$$r_i(0+) = \lim_{p \to -\infty} \frac{-p}{a_i(p)} = 0,$$

which follows from (84).

This implies that the limit, as $y = y_n \to 0$, of (136) can be taken. Therefore;

$$\begin{aligned} \frac{1}{c_i(z)} \mathbf{P}_{i,0} g(\mathfrak{n}_{\tau_z \wedge \mathfrak{s}}, \mathfrak{z}_{\tau_z \wedge \mathfrak{s}}) &= b_i^1 g'(i,0) + \int_0^\infty m_i(dy) g(i,y) + \sum_{j \neq i} \int_0^\infty m_{ij}(dy) g(j,y) \\ &+ b_i^1 h_z'(i,0) + \int_0^\infty m_i(dy) h(i,y), \end{aligned} \tag{138}$$

where $g(i,0) = 0$ and $g \in \mathbb{C}_2$. The inequality;

$$|h_z(i,x)| \leq \mathbf{P}_{i,x} \int_0^{\tau_1^0 \wedge \mathfrak{s}} |\mathfrak{A} g(i, \mathfrak{z}_t)| dt, \qquad 0 < z < 1,$$

implies that the limit, as $z \to 0$, of (138) can be taken. As a result, relation (133) remains valid as y approaches zero for any choice of y_n and for any $g \in \mathbb{C}_2$ (and hence for all continuous bounded differentiable, at the point $(i,0)$, functions, g).

In a similar way, the leaving time of $\mathbf{P}_{i,0} \tau_z \wedge \mathfrak{s}$ can be analysed. Let $0 < y < z < 1$ and $\mathfrak{s}_t = \inf\{t : \mathfrak{n}_t \neq i\}$. Then;

$$\mathbf{P}_{i,0} \tau_z \wedge \mathfrak{s} = \mathbf{P}_{i,0} \tau_z \wedge \mathfrak{s}_i = \mathbf{P}_{i,0} \tau_y \wedge \mathfrak{s}_i + \mathbf{P}_{i,0} \mathbf{P}_{\mathfrak{n}_{\tau_y \wedge \mathfrak{s}_i}, \mathfrak{z}_{\tau_y \wedge \mathfrak{s}_i}} (\tau_z \wedge \mathfrak{s}_i)$$

$$= \mathbf{P}_{i,0}\tau_y \wedge \mathfrak{s}_i + \mathbf{P}_{i,0}[\tau_y < \mathfrak{s}_i, \mathbf{P}_{i,\mathfrak{z}_{\tau_y}}(\tau_z \wedge \mathfrak{s}_i)]$$

and;

$$\begin{aligned} \mathbf{P}_{i,x}\tau_z \wedge \mathfrak{s}_i &= \mathbf{P}_{i,x}\tau_z^0 \wedge \mathfrak{s}_i + \mathbf{P}_{i,x}[\tau_z^0 < \mathfrak{s}_i, \mathbf{P}_{i,\mathfrak{z}_{\tau_z^0}}(\tau_z \wedge \mathfrak{s}_i)] \\ &= \mathbf{P}_{i,x}\tau_z^0 \wedge \mathfrak{s}_i + \mathbf{P}_{i,x}(\tau_z^0 < \mathfrak{s}_i, \mathfrak{z}_{\tau_z^0} = 0)\mathbf{P}_{i,0}\tau_z \wedge \mathfrak{s}_i. \end{aligned}$$

Further;

$$\frac{1}{c_j(z)}\mathbf{P}_{i,0}\tau_z \wedge \mathfrak{s} = \frac{1}{c_i(y)}\mathbf{P}_{i,0}\tau_y \wedge \mathfrak{s} + \frac{1}{c_i(y)}\mathbf{P}_{i,0}[\tau_y < \mathfrak{s}, \mathbf{P}_{i,\mathfrak{z}_{\tau_y}}(\tau_z^0 \wedge \mathfrak{s})]. \tag{139}$$

For $\mathfrak{A}g(i,y) \equiv 1$, it follows from (137) that:

$$\mathbf{P}_{i,x}\tau_z^0 \wedge \mathfrak{s} = \frac{r_i(z-x)}{r_i(z)}\int_0^z r_i(y)dy - \int_0^{z-x} r_i(y)dy,$$

where $x \in (0, z)$.

This implies the differentiability of $\mu_z(i,x) = \mathbf{P}_{i,x}\tau_z^0 \wedge \mathfrak{s}$ with respect to x at zero and the continuity of $\mu_z(i,x)$ in $x \geq 0$. Consequently, the limit, as $y \to 0$, of the second term in the right-hand side of (139) exists and is equal to:

$$b_i^1\mu_z(i,0) + \int_0^\infty m_i(dy)\mu_z(i,y).$$

Hence, the limit;

$$b_i^2 = \lim_{y\to 0}\frac{1}{c_i(y)}\mathbf{P}_{i,0}\tau_y \wedge \mathfrak{s} \tag{140}$$

exists and;

$$\frac{1}{c_i(z)}\mathbf{P}_{i,0}\tau_z \wedge \mathfrak{s} = b_i^2 + b_i^1\mu_z(i,0) + \int_0^\infty m_i(dy)\mu_z(i,y). \tag{141}$$

Let $b_i^2 > 0$. Then, as (118), (133), and (140) demonstrate, the domain of the operator, $\mathfrak{G}$, at the point, $(i,0)$, contains all functions, g, which are continuous, bounded, differentiable at the point $(i,0)$, and:

$$\begin{aligned} \mathfrak{G}g(i,0) &= \frac{1}{b_i^2}\left\{b_i^0 g(i,0) + \sum_{j\neq i}\int_0^\infty m_{ij}(dy)g(j,y)\right. \\ &\quad \left. + b_i^1 g'(i,0) + \int_0^\infty m_i(dy)[g(i,y) - g(i,0)]\right\}. \end{aligned} \tag{142}$$

If $b_i^2 = 0$, then the condition;

$$\begin{aligned} &b_i^0 g(i,0) + \sum_{j\neq i}\int_0^\infty m_{ij}(dy)g(i,y) + b_i^1 g'(i,0) \\ &\quad + \int_0^\infty m_i(dy)[g(i,y) - g(i,0)] = 0 \end{aligned} \tag{143}$$

is necessary for the existence of the limit given in (118) if the function g is bounded, continuous, and differentiable at zero. Moreover, if $g \in \mathbb{C}_2$, then this condition is sufficient for the limit given in (118) to exist and $\mathfrak{G}g(i,0) = \mathfrak{A}g(i,0)$.

Indeed, if (143) is valid, then (138) and (141) demonstrate that the fraction under the integral sign in (118) is:

$$\frac{b_i^1 h_z'(i,0) + \int_0^\infty m_i(dy) h_z(t,y)}{b_i^1 \mu_z'(i,0) + \int_0^\infty m_i(dy)\mu_z(i,y)}. \tag{144}$$

Let, for a given $\epsilon > 0$, the number $z > 0$ be chosen such that:

$$\mathfrak{A}g(i,0) - \epsilon \leq \mathfrak{A}g(i,y) \leq \mathfrak{A}g(i,0) + \epsilon$$

for $0 < y < z$. Then;

$$[\mathfrak{A}g(i,0) - \epsilon]\mu_z(i,y) \leq h_z(i,y) \leq [\mathfrak{A}g(i,0) + \epsilon]\mu_z(i,y). \tag{145}$$

If this inequality is divided by y and the limit as $y \downarrow 0$ is taken, this inequality becomes:

$$[\mathfrak{A}g(i,0) - \epsilon]\mu_z'(i,0) \leq h_z'(i,0) \leq [\mathfrak{A}g(i,0) + \epsilon]\mu_z'(i,0). \tag{146}$$

If inequality (145) is integrated with respect to $m_i(dy)$ and the result is added to inequality (146), multiplied by b_i^1, then, as a result, the fraction given in (144) is demonstrated to be bounded from the left by $\mathfrak{A}g(i,0) - \epsilon$ and from the right by $\mathfrak{A}g(i,0) + \epsilon$, i.e. limit (118) is equal to $\mathfrak{A}g(i,0)$.

Finally, if the state, $(i,0)$, is absorbing, then $\mathfrak{G}g(i,0) = 0$ and, consequently, equation (142) is valid with $b_i^2 = 1$ and $b_i^0 = b_i^1 = m_i(0,\infty) = m_{ij}[0,\infty) = 0$.

The coefficients b_i^0, b_i^1, b_i^2, $m_i(dy)$ and $m_{ij}(dy)$, constructed in this manner, satisfy conditions (115) and (116). All the functions from $\mathbb{D} \cap \mathbb{C}_2$, and only these, satisfy the boundary-value condition given in (114).

To finish the proof, it remains to verify that condition (117) is valid. This can be demonstrated by contradiction. Suppose it is not valid, i.e. $b_i^1 = b_i^2 = 0$ and $m_i(0,\infty) < \infty$. Then, it follows from (131) and (132) that:

$$\frac{1}{c_i(y)} \underset{y\to\infty}{\longrightarrow} -b_i^0 + m_i(0,\infty) < \infty. \tag{147}$$

Hence;

$$b_i^0 g(i,0) + \sum_{j\neq i} \int_0^\infty m_{ij}(dy) g(j,y) + \int_0^\infty m_i(dy)[g(i,y) - g(i,0)] = 0$$

is a necessary condition for the existence of the limit given in (118) in the class $\mathbb{C}$. Because the characteristic operator is an extension of the infinitesimal operator, this contradicts the fact that the domain of the infinitesimal operator is dense. This contradiction demonstrates that condition (117) is valid and thus the theorem is proved.

57. The following statement is the converse to Theorem 56.

Theorem. *If the coefficients of the boundary-value condition (114) satisfy conditions (115), (116) and (117). Then, the boundary-value condition (114) defines the domain, in* $\mathbb{C}_2$, *of the infinitesimal operator of a random walk on the semiaxis.*

Proof. The subset, $\mathbb{D}_2$, of $\mathbb{C}_2$, which is defined by condition (114), is dense in $\mathbb{C}$ (in the uniform topology). Indeed, let $\mathrm{l}(f)$ be the difference between the right and the left hand sides of equation (114), and the set $\mathbb{C}_2$ has the structure of a linear normed space with the norm $\|f\| = \max_{i,x} |f(i,x)|$.

Conditions (84) and (117) imply that the mapping from $\mathbb{C}_2$ into $\mathbf{R}^d$, defined by $f \to \mathrm{l}(f)$, is unbounded. Consequently, the image of the unit solid sphere in $\mathbb{C}_2$ under the mapping l is a convex symmetric unbounded set in $\mathbf{R}^d$. Moreover, this set coincides with $\mathbf{R}^d$ because the linear functional, $f \to \langle \mathbf{v}, \mathrm{l}(f)\rangle$, is unbounded for any nonzero vector, $\mathbf{v}$, from $\mathbf{R}^d$. Consequently, the image of any solid sphere from $\mathbb{C}_2$, of positive radius, under the mapping l is $\mathbf{R}^d$. However, each sphere has a nonempty intersection with the set $\mathbb{D}_2 = \{f \in \mathbb{C}_2 : \mathrm{l}(f) = 0\}$, which implies the density of the latter in $\mathbb{C}_2$ and, hence, in $\mathbb{C}$.

The operator $\mathfrak{A}$ satisfies the following *maximum principle* on $\mathbb{D}_2$: if a function f from $\mathbb{D}_2$ achieves its maximum at a point (i,x), then $\mathfrak{A}f(i,x) \le 0$. Indeed, if $x > 0$ and $f(i,x) = \max_{j,y} f(j,y)$, then $f'(i,x) = 0$, $f''(i,x) \le 0$, and consequently;

$$\begin{aligned}\mathfrak{A}f(i,x) &= \sum_{j\neq i} \int_0^\infty l_{ij}(dy)[f(j,x+y) - f(i,x)] + a_i^2 f''(i,x) \\ &\quad + \int_0^\infty l_i(dy)[f(i,x+y) - f(i,x)] \le 0. \qquad (148)\end{aligned}$$

If $f(i,0) = \max_{j,y} f(j,y)$, then $f'(i,0) \le 0$ and, in the case where $b_i^2 > 0$;

$$\begin{aligned}\mathfrak{A}f(i,0) &= \frac{1}{b_i^2}\left\{\sum_{j\neq i} \int_0^\infty m_{ij}(dy)[f(j,y) - f(i,0)] + b_i^1 f'(i,0)\right. \\ &\quad \left. + \int_0^\infty m_i(dy)[f(i,y) - f(i,0)]\right\} \le 0.\end{aligned}$$

If $b_i^2 = 0$, the condition $f \in \mathbb{D}_2$ implies that the expression in brackets is equal to zero. This is only possible if:

$$b_i^1 f'(i,0) + \int_0^\infty m_i(dy)[f(i,y) - f(i,0)] = 0.$$

This, together with (117), implies that $f'(i,0) = 0$. Hence, $f''(i,0) \le 0$ and inequality (148) is valid at $x = 0$.

This maximum principle implies the uniqueness, in $\mathbb{D}_2$, of the trivial solution of the equation;

$$\mathfrak{A}g(i,x) + sg(i,x) = 0, \qquad i = 1, \dots, d, \quad x \ge 0 \qquad (149)$$

for $s < 0$. Indeed, let $g \in \mathbb{D}_2$ and (149) be valid. Then, the function, g, reaches a maximum at some point, (t^*, x^*), and, consequently;

$$\max_{j,y} g(j,y) = g(i^*, x^*) = -\frac{1}{s}\mathfrak{A}g(i^*, x^*) \leq 0.$$

However, the function, $-g$, also belongs to $\mathbb{D}_2$ and satisfies equation (149). Thus; $\min_{j,y} g(j,y) = -\max_{j,y}[-g(j,y)] \geq 0$, i.e. $g \equiv 0$.

Let $s < 0$ be chosen in such a way that, for any integer $n \leq 0$, the matrix $A(n) + sE$ is invertible. Then, the equation;

$$\mathfrak{A}g(i,x) + sg(i,x) = e^{nx}q_i, \qquad i = 1, \dots, d, \quad x \geq 0, \tag{150}$$

has a solution in $\mathbb{D}_2$ for all integers $n \leq 0$. It is unique because the trivial solution of (149) is unique in $\mathbb{D}_2$. It is convenient to look for a solution of (150) in the form:

$$\mathbf{g}(x) = e^{xQ(s)}\mathbf{u} + e^{nx}\mathbf{v}, \tag{151}$$

where $\mathbf{u}, \mathbf{v} \in \mathbf{R}^d$, $\mathbf{g}(x) = (g(i,x))_{i=1}^d$ and the matrix $Q(s)$ is a solution of equation (92).

It follows from (89), (92), and (151) that;

$$\mathfrak{A}\mathbf{g}(x) + s\mathbf{g}(x) = e^{nx}[A(n) + sE]\mathbf{v} = e^{nx}\mathbf{q},$$

where $\mathbf{q} = (q_i)_{i=1}^d$. Hence, if $\mathbf{v} = [A(n) + sE]^{-1}\mathbf{q}$, then the function given in (151) satisfies relation (150) for any vector, $\mathbf{u}$. The vector, $\mathbf{u}$, is defined from the condition $\mathbf{g} \in \mathbb{D}_2$. In this notation;

$$B(p) = B_0 + \int_0^\infty M_0(dy)e^{py} + B_1 p + \int_0^\infty M_1(dy)(e^{py} - 1), \qquad \Re p \leq 0.$$

The condition $g \in \mathbb{D}_2$ for the function (151) with a chosen vector, $\mathbf{v}$, is reduced to the form:

$$[B(Q(s)) + sB_2]\mathbf{u} = [B(n) + sB_2]\mathbf{v} + B_2\mathbf{q},$$

where;

$$B(Q) = B_0 + \int_0^\infty M_0(dy)e^{yq} + B_1 Q + \int_0^\infty M_1(dy)(e^{yQ} - E).$$

The non-diagonal elements of the matrix $B(s)$ are nonnegative, and the sum of the elements in any given row, by (115), are nonpositive. The matrix $B(Q(s))$ also possesses this property. Moreover, it follows from condition (117) that, for $s < 0$, the sum of the elements in any given row of the matrix $B(Q(s)) + sB_2$ are negative. Consequently, this matrix is invertible, and this implies that the vector, $\mathbf{u}$, can be determined from the last equation.

Let $\mathbb{F}$ be a subset of $\mathbb{C}$ consisting of finite linear combinations of the functions $f(i,x) = e^{-kx}q_i$ ($k = 0, 1, 2, \dots$, $q_i \in \mathbf{R}^1$). The set, $\mathbb{F}$, is dense in $\mathbb{C}$. This follows from the uniform approximation of continuous functions by Bernstein's polynomials. This also implies that a nonnegative function from $\mathbb{C}$ is a uniform limit of nonnegative functions from $\mathbb{F}$.

Thus, for a dense, in $\mathbb{C}$, set of functions, f, the equation;

$$\mathfrak{A}g + sg = f \tag{152}$$

has, in $\mathbb{D}$, a unique solution for all $s < 0$ except at most, a countable number of s. From the maximum principle, it follows that these solutions are nonpositive if the right hand side is nonnegative. Indeed, if $g(i^*, x^*) = \max g(j, y)$, then;

$$\max g(j,y) = \frac{1}{s} f(i^*, x^*) - \frac{1}{s}\mathfrak{A}g(i^*, x^*) \le \frac{1}{s} f(i^*, x^*). \tag{153}$$

The right hand side of this inequality is nonpositive for $f \ge 0$. Hence, equation (152) defines a positive operator, $\mathfrak{R}_s$, acting from $\mathbb{F}$ into $\mathbb{D}_2$ by the formula $\mathfrak{R}_s f = -g$. It also follows from inequality (153) that the operator $\mathfrak{R}_s$ is bounded and its norm does not exceed $1/|s|$. Moreover, the norm of $\mathfrak{R}_s$ is precisely $1/|s|$.

Indeed, if $g(i,x) \equiv 1$, then $g \in \mathbb{D}_2$ and $\mathfrak{A}g(i,x) \equiv 0$. So, if $f(i,x) \equiv 1$, then $\mathfrak{R}_s f(i,x) \equiv -1/s$.

By continuity, the operator $\mathfrak{R}_s$ can be extended to $\mathbb{C}$ while preserving its norm. For values of s_1 and s_2 for which equation (152) has a solution, the following resolvent equation is valid:

$$\mathfrak{R}_{s_1} - \mathfrak{R}_{s_2} = (s_1 - s_1)\mathfrak{R}_{s_1}\mathfrak{R}_{s_2}. \tag{154}$$

To demonstrate this, let $f \in \mathbb{F}$, $g_1 = \mathfrak{R}_{s_1} f$ and $g_2 = \mathfrak{R}_{s_2} f$. Then, $g_1, g_2 \in \mathbb{D}_2$, $\mathfrak{A}g_1 + s_1 g_1 = \mathfrak{A}g_2 + s_2 g_2$ or $\mathfrak{A}(g_1 - g_2) + s_1(g_1 - g_2) = -(s_1 - s_2)g_2$. Because $g_1 - g_2 \in \mathbb{D}_2$, the last equation implies (154).

The resolvent equation implies that the operator, $\mathfrak{R}_s$, is uniformly continuous in $s \le c$ for any $c < 0$. This implies that $\mathfrak{R}_s$ can be defined for all values of $s < 0$ such that the resolvent equation is valid, which demonstrates that the range of the operator, $\mathfrak{R}_s$, does not depend on s and its closure is equal to $\mathbb{C}$. This last assertion follows from the density of $\mathbb{D}_2$ in $\mathbb{C}$ and the inclusion $\mathfrak{R}_s(\mathbb{C}) \supset \mathbb{D}_2$.

Thus the constructed family of operators, $\mathfrak{R}_s$ $(s < 0)$, satisfies conditions of the Hille-Iosida theorem, which, together with Kolmogorov's theorem, yields the existence and uniqueness (with respect to the distribution) of a homogeneous Markovian Feller process, $(\mathfrak{n}_t(\mathfrak{z}_t),\ t \ge 0)$, in the phase space $\{1, \ldots, d\} \times [0, \infty)$ such that;

$$\mathfrak{R}_s f(i,x) = \mathbf{P}_{i,x} \int_0^\infty e^{st} f(\mathfrak{n}_t, \mathfrak{z}_t) dt.$$

Moreover, since the operator ,$\mathfrak{R}_s$, maps the space of continuous functions vanishing at infinity into itself, the trajectories of the process, $(\mathfrak{n}_t, \mathfrak{z}_t)$ $(t \ge 0)$, can be chosen to be right continuous and, at every fixed stopping time, to be almost certainly left continuous, and the set of all these trajectories can be considered as closed with respect to translation and the stopping times.

To finish the proof, it remains to verify that the distributions of the processes (ν_t, ζ_t) $(t < \theta)$ and $(\mathfrak{n}_t, \mathfrak{z}_t)$ $(t < \tau$, where $\tau = \inf\{t : \mathfrak{z}_t = 0\})$ coincide.

First of all, the process $\mathfrak{z}_t$ $(t \geq 0)$ will be demonstrated not to have negative jumps. Let τ^z be the moment at which the process $\mathfrak{z}_t$ $(t \geq 0)$ leaves the interval (z, ∞) for the first time and let a function, f, from $\mathbb{D}_2$ be equal to zero for all values of (i, x) with $x > z$. By Dynkin's formula;

$$\mathbf{P}_{i,x} e^{s\tau^z} f(\mathfrak{n}_{\tau^z}, \mathfrak{z}_{\tau^z}) = f(i, x) + \mathbf{P}_{i,x} \int_0^{\tau^z} e^{st} g(\mathfrak{n}_t, \mathfrak{z}_t) dt, \tag{155}$$

where $g(j, y) = \mathfrak{A} f(j, y) + s f(j, y)$. Because $f(j, y) = 0$ for $y > z$, then $g(j, y) = 0$ for $y > z$. However, $\mathfrak{z}_t > z$ for $t < \tau^z$. Consequently, the right hand side of (155) is equal to zero for $x > z$ and, because $\mathbb{D}_2$ is dense in $\mathbb{C}$, $\mathfrak{z}_{\tau^z}$ is almost certainly (with respect to $\mathbf{P}_{i,x}$) equal to z. Since there are no discontinuities of the second kind then there are no negative jumps which implies that, for $x > 0$, $\mathbb{C}_2$ is a subset of the domain of the characteristic operator, $\mathfrak{G}$, of the process $(\mathfrak{n}_t, \mathfrak{z}_t)$ $(t \geq 0)$ at the point (i, x) and that the equation;

$$\mathfrak{G} f(i, x) = \mathfrak{A} f(i, x) \tag{156}$$

is valid for $x > 0$ and $f \in \mathbb{C}_2$.

Indeed, because the process $\mathfrak{z}_t$ $(t \geq 0)$ does not have negative jumps, the ratio;

$$\frac{\mathbf{P}_{i,x} g(\mathfrak{n}_{\tau_\epsilon^\delta(x) \wedge \mathfrak{s}}, \mathfrak{z}_{\tau_\epsilon^\delta(x) \wedge \mathfrak{s}}) - g(i, x)}{\mathbf{P}_{i,x} \tau_\epsilon^\delta(x) \wedge \mathfrak{s}}, \tag{157}$$

where $x > 0$, $\tau_\epsilon^\delta(x) = \inf\{t : \mathfrak{z}_t \notin (x - \epsilon, x + \delta)\}$ and $\mathfrak{s} = \inf\{t : \nu_t - \nu_0\}$, does not depend on the values, $g(j, y)$, for $0 \leq y < x - \epsilon$.

Moreover, for any function, g, from $\mathbb{C}_2$, there exists a function, f, from $\mathbb{D}_2$ such that $g(j, y) = f(j, y)$ for $y \geq x - c_i v(i, y)$, where $v(i, y)$ is a smooth finite function with the support contained in the interval $[0, x - \epsilon)$ for every $i = 1, \ldots, d$ and $c_i = l_i(g)/l_i(v)$, where $l_i(f)$ is the ith coordinate of the vector $\mathbf{l}(f)$, which is equal to the difference of the left and the right hand sides of the boundary-value condition (114) and (117) implies that it is possible to choose the function, v, satisfying $l_i(v) \neq 0$ $(i = 1, \ldots, d)$. Thus, if there is a function, g, from $\mathbb{C}_2$ in (157), then it can be assumed that $g \in \mathbb{D}_2$. The observation that a characteristic operator is an extension of the infinitesimal operator completes the proof.

The next step is to prove the formula:

$$\|\mathbf{P}_{i,x}(e^{s\tau}, \mathfrak{n}_\tau = j)\|_{i,j=1}^d = e^{xQ(s)}, \tag{158}$$

where the matrix $Q(s)$ is defined by (87).

If $f \in \mathbb{C}$ and $f(j,y) = 0$ for $y \geq \epsilon > 0$, then;

$$\begin{aligned}\mathfrak{R}_s f(i,x) &= \mathbf{P}_{i,x} \int_0^{\tau^\epsilon} e^{st} f(\mathfrak{n}_t, \mathfrak{z}_t)dt + \sum_j \mathbf{P}_{i,x}(e^{s\tau^\epsilon}, \mathfrak{n}_{\tau^\epsilon} = j)\mathfrak{R}_s f(j,\epsilon) \\ &= \sum_j \mathbf{P}_{i,x}(e^{s\tau^\epsilon}, \mathfrak{n}_{\tau^\epsilon} = j)\mathfrak{R}_s f(j,\epsilon).\end{aligned}$$

The vectors $\mathfrak{R}_s \mathbf{f}(\epsilon) = (\mathfrak{R}_s f(j,\epsilon))_{j=1}^d$ forms an invertible matrix. Hence, the matrix $U_\epsilon(x) = \|\mathbf{P}_{i,x}(e^{s\tau^\epsilon}, \mathfrak{n}_{\tau^\epsilon} = j)\|_{i,j=1}^d$ continuously depends on x and:

$$\lim_{x \to \infty} U_\epsilon(x) = 0.$$

Moreover, $U_\epsilon(x)$ is a solution of the equation:

$$\mathfrak{G} U_\epsilon(x) + sU_\epsilon(x) = 0, \qquad x > \epsilon, \quad U_\epsilon(\epsilon) = E. \tag{159}$$

By (89), (92), and (156), the matrix $e^{(x-\epsilon)Q(s)}$ is a solution of this equation. However, equation (159) can not have more than one solution for each $s < 0$ in the class of the matrices, which are continuous in $x \geq \epsilon$ and vanish for $x \to \infty$. Consequently, $U_\epsilon(x) = e^{(x-\epsilon)}Q(s)$ for $x \geq \epsilon$ and $s < 0$.

Because the process $(\mathfrak{n}_t, \mathfrak{z}_t)$ $(t \geq 0)$ is almost certainly left continuous at every stopping time, $\tau = \lim_{\epsilon \downarrow 0} \tau_\epsilon$ is almost certainly true with respect to $\mathbf{P}_{i,x}$. This proves (158), which, in particular, demonstrates that:

$$\mathbf{P}_{i,x}(e^{s\tau}, \mathfrak{n}_\tau = j) = \mathbf{P}_{i,x}(e^{s\theta}, \nu_\theta = j). \tag{160}$$

Let $f \in \mathbb{D}_2$. By Dynkin's formula and from (156);

$$\begin{aligned}\mathbf{P}_{i,x} e^{s\tau} f(\mathfrak{n}_\tau, 0) &= f(i,x) + \mathbf{P}_{i,x} \int_0^\tau e^{st} g(\mathfrak{n}_t, \mathfrak{z}_t)dt, \\ \mathbf{P}_{i,x} e^{s\theta} f(\nu_\theta, 0) &= f(i,x) + \mathbf{P}_{i,x} \int_0^\theta e^{st} g(\nu_t, \zeta_t)dt,\end{aligned}$$

where $g(j,y) = \mathfrak{A} f(j,y) + sF(j,y)$. This, together with (160) implies that:

$$\mathbf{P}_{i,x} \int_0^\tau e^{st} g(\mathfrak{n}_t, \mathfrak{z}_t)dt = \mathbf{P}_{i,x} \int_0^\theta e^{st} g(\nu_t, \zeta_t)dt$$

for a dense, in $\mathbb{C}$, set of functions, g (which is the image of the set $\mathbb{D}_2$ under the mapping $\mathfrak{A} + sJ$). This proves that the distributions of the processes $(\mathfrak{n}_t, \mathfrak{z}_t)$ $(t < \tau)$ and (ν_t, ζ_t) $(t < \theta)$ coincide. The theorem is thus proved.

58. In the course of the proof of Theorem 57 it was shown that equation (152) has a solution belonging to $\mathbb{D}_2$ if the function in the right hand side of the equation belongs to a dense subset of $\mathbb{C}$. This demonstrates that restricting the domain of the infinitesimal operator of the process $(\mathfrak{n}_t, \mathfrak{z}_t)$ $(t \geq 0)$ from $\mathbb{D}$ to $\mathbb{D}_2 = \mathbb{D} \cap \mathbb{C}_2$, does not lead to a loss of information about the process and, hence, the statement of Theorem 56 is informative.

59. Condition (84) guarantees the existence of a unique (up to a common diagonal factor) way of representing the set $\mathbb{D}_2 = \mathbb{D} \cap \mathbb{C}_2$ by equation (114). In more precise terms, let $\overline{\mathbb{D}}_2$ be a subset of $\mathbb{C}_2$ defined by the boundary-value condition;

$$\overline{B}_2\mathfrak{A}\mathbf{f}(0) = \overline{B}_0\mathbf{f}(0) + \int_0^\infty \overline{M}_0(dy)\mathbf{f}(y) + \overline{B}_1\mathbf{f}'(0) + \int_0^\infty \overline{M}_1(dy)[\mathbf{f}(y) - \mathbf{f}(0)],$$

the coefficients of which satisfy conditions of the type given in (115), (116) and (117). Then, if $\mathbb{D}_2 \subset \overline{\mathbb{D}}_2$, then $\mathbb{D}_2 = \overline{\mathbb{D}}_2$, $\overline{B}_2 = CB_0$, $\overline{B}_1 = CB_1$, $\overline{B}_2 = CB_2$, $\overline{M}_0(dy) = CM_0(dy)$ and $\overline{M}_1(dy) = CM_1(dy)$, where C is a diagonal matrix with positive elements on the main diagonal.

Indeed, the function $\mathbf{f}(x) = e^{qx}\mathbf{u} + e^{px}\mathbf{v}$ $(x \geq 0)$ is an element of $\mathbb{D}_2$ if and only if:

$$[B_2A(q) - B(q)]\mathbf{u} = [B_2A(p) - B(p)]\mathbf{v}.$$

Consequently, the inclusion $\mathbb{D}_2 \subset \overline{\mathbb{D}}_2$ implies (in natural notation) that:

$$[B_2A(q) - B(q)]^{-1}[B_2A(p) - B(p)] = [\overline{B}_2A(q) - \overline{B}(q)]^{-1}[\overline{B}_2A(p) - \overline{B}(p)]$$

which can be written as:

$$\sum_j c_{ij}[b_j^2 a_{jk}(p) - b_{jk}(p)] = \overline{b}_i^2 a_{ik}(p) - \overline{b}_{ik}(p),$$

where $\|c_{ij}\|_{i,j=1}^d = C = [\overline{B}_2A(q) - \overline{B}(q)][B_2A(q) - B(q)]^{-1}$. Conditions (84) and (117) show that;

$$b_k^2 a_{kk}(p) - b_{kk}(p) \to \infty \qquad \text{as} \qquad p \to -\infty.$$

Consequently, the matrix C is diagonal and its diagonal elements, $c_i = c_{ii}$, are positive and:

$$c_j[b_j^2 a_{ij}(p) - b_{ij}(p)] = \overline{b}_i^2 a_{ij}(p) - \overline{b}_{ij}(p).$$

Because, by (84) and (116), $a_{jj}(p)$ converges faster than $b_{jj}(p)$ as $p \to -\infty$, also, $c_j b_j^2 = \overline{b}_j^2$ and consequently, $c_j b_{ij}(p) = \overline{b}_{ij}(p)$ which completes the proof.

60. The following simple, but useful in this book, lemma follows from relations (133) and (140) and the definition given in (118) of the characteristic operator, $\mathfrak{G}$.

Lemma. *If a continuous bounded function g, which is differentiable at the point $(i,0)$, belongs to the domain of the operator $\mathfrak{G}$ at the point $(i,0)$, then;*

$$\begin{aligned} b_i^2\mathfrak{G}g(i,0) &= b_i^0 g(i,0) + \sum_{j\neq i}\int_0^\infty m_{ij}(dy)g(j,y) + b_i^1 g'(i,0) \\ &\quad + \int_0^\infty m_i(dy)[g(i,y) - g(i,0)]. \end{aligned}$$

3.3 The ergodic theorem

61. The ergodic properties of a random walk on a semiaxis are nontrivial if and only if the walk is not absorbed in any of the states of the form $(i,0)$. In terms of the boundary condition given in (114), the absorption in a set of states $(i,0)$ implies that there exists a set of indices, $\mathcal{I} \subset \{1,\dots,d\}$, such that;

$$\begin{aligned} m_{ij}[0,\infty) = 0 \qquad &\text{for} \quad i \in \mathcal{I}, \quad j \notin \mathcal{I}, \\ b_i^1 = m_i(0,\infty) = m_{ij}(0,\infty) = 0 \qquad &\text{for} \quad i \in \mathcal{I}, \quad j \in \mathcal{I}. \end{aligned} \tag{161}$$

Indeed, if $b_i^1 = 0$ and $m_i(0,\infty) < \infty$ then it follows from (147) and (133) that there exists a positive limit $\tau_{0+} = \lim_{y\downarrow 0} \tau_y$ and, if $m_i(0,\infty) > b_i^0$, then;

$$\mathbf{P}_{i,-}(\mathfrak{n}_{\tau_{0+}\wedge s} = j, \mathfrak{z}_{\tau_{0+}\wedge s} \geq y) = \begin{cases} \dfrac{m_{ij}[y,\infty)}{-b_i^0 + m_i(0,\infty)} & \text{for} \quad i \neq j, \\ \dfrac{m_i[y,\infty)}{-b_i^0 + m_i(0,\infty)} & \text{for} \quad i = j. \end{cases} \tag{162}$$

The random variable $\tau_{0+} \wedge s$ almost certainly (with respect to $\mathbf{P}_{i,0}$) coincides with the time that the process first leaves the state $(i,0)$, and (162) demonstrates that:

$$\mathfrak{n}_{\tau_{0+}\wedge s} \in \mathcal{I}$$

is almost certainly true with respect to $\mathbf{P}_{i,0}$ for $i \in \mathcal{I}$.

Hence, after the process $(\mathfrak{n}_t, \mathfrak{z}_t)$ $(t \geq 0)$ reaches the set of states in $\mathcal{I} \times \{0\}$ defined by the conditions given in (161), it remains there and degenerates into a Markovian process with a finite number of states.

Hence, without loss of generality, it can be assumed that the conditions given in (161) are valid only in the trivial situation;

$$\mathcal{I} = \emptyset. \tag{163}$$

62. If the conditions given in (163) are valid, then the ergodic properties of the process $(\mathfrak{n}_t, \mathfrak{z}_t)$ $(t \geq 0)$ are completely determined by the sign of the number;

$$\mu = \sum_i \rho \left[a_i^1 + \int_1^\infty (y-1) l_i(dy) + \sum_{i \neq j} \int_0^\infty y l_{ij}(dy) \right],$$

where $(\rho_i)_{i=1}^d = \rho$ is the vector of the stationary probabilities of the Markovian process ν_t $(t \geq 0)$, of which all states are connectable, i.e, $A(0)^\top \rho = \mathbf{0}$ and $\langle \rho, \mathbf{1} \rangle = 1$.

When $\mu \leq 0$ then the process, $(\mathfrak{n}_t, \mathfrak{z}_t)$ $(t \geq 0)$, possesses an invariant measure expressed in terms of both the elements of the matrix;

$$\mathfrak{B}R(x) \;=\; B_0 R(x) + \int_0^\infty M_0(dy) R(x-y) - B_1 R'(x)$$

$$+\int_0^\infty M_1(dy)[R(x-y)-R(x)], \tag{164}$$

where $R(x) = R_0(x)$, $R'(x) = \lim_{y\downarrow 0}\frac{1}{y}[R(x)-R(x-y)]$ and $R(y)=0$ for $y \le 0$, and the components, e_i $(i=1,\ldots,d)$, of the left annulling vector, $\mathbf{e}$, of the matrix;

$$B(Q) = B_0 + \int_0^\infty M_0(dy)e^{yQ} + B_1 Q + \int_0^\infty M_1(dy)[e^{yQ}-E],$$

where $Q = Q(0-)$. The atom of the invariant measure at the point $(i,0)$ is equal to $b_i^2 e_i$ and, when the second component is not equal to zero, the densities, $\pi_i(x)$ $(i=1,\ldots,d)$, of the invariant measure coincide with the coordinates of the vector $-\mathfrak{B}R(x)^\top\mathbf{e}$.

Theorem. *Let $\mu \le 0$. If:*

$$\sum_{i=1}^d \int_0^\infty \{|f(i,x)| + |g(i,x)|\}\pi_i(x)dx < 0$$

and:

$$\sum_{i=1}^d \left\{g(i,0)b_i^2 e_i + \int_0^\infty g(i,x)\pi_i(x)dx\right\} \neq 0,$$

then:

$$\lim_{t\to\infty}\frac{\int_0^t f(\mathfrak{n}_t,\mathfrak{z}_t)dt}{\int_0^t g(\mathfrak{n}_t,\mathfrak{z}_t)dt} = \frac{\sum_{i=1}^d\left\{f(i,0)b_i^2e_i + \int_0^\infty f(i,x)\pi_i(x)dx\right\}}{\sum_{i=1}^d\left\{g(i,0)b_i^2e_i + \int_0^\infty g(i,x)\pi_i(x)dx\right\}} \tag{165}$$

is almost certainly true with respect to $\mathbf{P}_{i,x}$, where $(e_i)_{i=1}^d = \mathbf{e}$, $B(Q)^\top\mathbf{e} = 0$, $\langle\mathbf{e},\mathbf{1}\rangle = 1$ and $(\pi_i(x))_{i=1}^d = -\mathfrak{B}R(x)^\top\mathbf{e}$.

Proof. Let, for a fixed $z > 0$, $\mathfrak{m}$ denote the time at which the trajectory of the process, $\mathfrak{z}_t$ $(t \ge 0)$, reaches zero for the first time after it leaves $[0,z)$, i.e.

$$\mathfrak{m} = \tau_z + T_{\tau_z}\tau,$$

where $\tau_z = \inf\{t : \mathfrak{z}_t \ge z\}$, $\tau = \inf\{t : \mathfrak{z}_t = 0\}$ and T_u is the operator of translation, by u, of the trajectory of the process $(\mathfrak{n}_t,\mathfrak{z}_t)$ $(t \ge 0)$.

For such a choice of the time, $\mathfrak{m}$, then the conditions of Theorem 40 will be valid. Indeed, by the complete probability formula;

$$\mathbf{P}_{i,x}(e^{s\mathfrak{m}},\mathfrak{n}_\mathfrak{m} = j) = \mathbf{P}_{i,x}e^{s\tau_z}\varphi_j(\mathfrak{n}_{\tau_z},\mathfrak{z}_{\tau_z}), \tag{166}$$

where $\varphi_j(i,y) = \mathbf{P}_{i,y}(e^{s\tau},\mathfrak{n}_\tau = j)$ and, consequently;

$$\|\varphi_j(i,y)\|_{i,j=1}^d = e^{yQ(s)}.$$

63. Lemma. *If $g \in \mathbf{C}^2$, then:*

$$(\mathbf{P}_{i,x}e^{s\mathcal{T}_z}g(\mathfrak{n}_{\mathcal{T}_z},\mathfrak{z}_{\mathcal{T}_z}))_{i=1}^d = g(x) + R_s(z-x)\mathbf{c} - \\ - \int_x^s R_s(y-x)[\mathfrak{A}\mathbf{g}(x) + s\mathbf{g}(x)]dx, \qquad (167)$$

where $x \in [0, x)$, $s \leq 0$, $\mathbf{g}(x) = (g(i,x))_{i=1}^d$ and the vector $\mathbf{c} = c(s,g,z)$ does not depend on x.

Proof. For $g \in \mathbb{D}_2$, the Lemma follows from Dynkin's formulae and the following formula:

$$\left(\mathbf{P}_{i,x}\int_0^{\mathcal{T}_z} e^{st}f(\mathfrak{n}_t,\mathfrak{z}_t)dt\right)_{i=1}^d = R_z(z-x)\mathbf{c} - \int_x^z R_s(y-x)\mathbf{f}dy, \qquad (168)$$

where $x \in [0,z)$, $s \leq 0$, $f \in \mathbf{C}$, $f(y) = (f(i,y))_{i=1}^d$ and the vector $\mathbf{c}$ does not depend on x.

Formula (168) is implied by (110), (111), and the following representation:

$$\mathbf{P}_{i,x}\int_0^{\mathcal{T}_z} e^{st}f(\mathfrak{n}_t,\mathfrak{z}_t)dt = \mathbf{P}_{i,x}\int_0^{\theta_z^0} e^{st}f(\nu_t,\zeta_t)dt \\ + \sum_{j=1}^d \mathbf{P}_{i,x}(e^{s\theta_z^0}, \zeta_{\theta_z^0} = 0, \nu_{\theta_z^0} = j)\mathbf{P}_{j,0}\int_0^{\mathcal{T}_z} e^{st}f(\mathfrak{n}_t,\mathfrak{z}_t)dt.$$

In the general case, for any function, $g \in \mathbb{C}_2$, and any $\epsilon > 0$, one can find a function, g_ϵ, from $\mathbb{D}_2$ which coincides with g for all values of the argument (i,x) with $x > \epsilon$. Moreover, it follows from $g(i,x) = g_\epsilon(i,x)$ for $x > \epsilon$ that $\mathfrak{A}g(i,x) = \mathfrak{A}g_\epsilon(i,x)$ for $x > \epsilon$.

Because the left hand side of (167) does not change if g is replaced by g_ϵ, with $\epsilon < z$ and $g_\epsilon \in \mathbb{D}_2$, then (167) is valid if $x > \epsilon$ for a vector $\mathbf{c} = \mathbf{c}_\epsilon$. The comparision of the right hand sides given by the formula (167), when ϵ takes different values, and also because the matrix $R_s(x)$ is invertible, demonstrates that the vector $\mathbf{c}$ does not depend on ϵ.

By using (89), (92) and applying Lemma 63 to the right hand side of (166), the following is obtained:

$$\|\mathbf{P}_{i,x}(e^{s\mathfrak{m}}, \mathfrak{n}_{\mathfrak{m}} = j)\|_{i,j=1}^d = e^{xQ(s)} + R_s(z-x)C. \qquad (169)$$

If Lemma 60 is used and because the operator $\mathfrak{G} + s\mathcal{I}$ in (169) is equal to zero, then the equation for determining the matrix, C, is obtained:

$$[\mathfrak{B}R_s(z) + sB_2R_s(z)]C = -B(Q(s)) - sB_2, \qquad (170)$$

where the matrix $\mathfrak{B}R_s(z)$ is defined by (164) with $R(x)$ replaced by $R_s(x)$. The matrix in the brackets is invertible for all $s \leq 0$.

Indeed;

$$[\mathfrak{B}R_s(z)+sB_2R_s(z)]R_s(z)^{-1}=B_0+sB_2-M_1[z,\infty)$$
$$+\int_0^z M_0(dy)R_s(z-y)R_s(z)^{-1}+\int_0^z M_1(dy)[R_s(z-y)R_s(z)^{-1}-E].$$

It follows from (110) and condition (84) that the matrix $R_s(z-y)R_s(z)^{-1}$ has, for $s\le 0$ and $0<y<z$, positive elements such that the sum of the elements in each row is less than unity. Consequently, the matrix in the right hand side of the last equation has nonnegative off-diagonal elements, and the sums of the elements in each row, by (117) and (163), are negative for all $s\le 0$. This proves that it is invertible and, hence, the matrix in the brackets in the left hand side of (170) is also invertible.

Hence, the matrix in the left hand side of (169) is equal to:

$$e^{xQ(s)}-R_s(z-x)\{\mathfrak{B}R_s(z)+sB_2R_s(z)\}^{-1}[B(Q(s))+sB_2]. \tag{171}$$

Consequently, $\mathbf{P}_{i,x}(\mathfrak{m}<\infty)=1$ is equivalent to $Q\mathbf{1}=\mathbf{0}$.

64. Lemma. *If $\mu\le 0$, then $Q\mathbf{1}=\mathbf{0}$.*

Proof. Let $\mathfrak{t}_1^i$, $\mathfrak{t}_2^i$, ... be a sequence of times at which the process, ν_t $(t\ge 0)$, returns to the state i. Then, the random variables $\zeta_{\mathfrak{t}_1^i}$, $\zeta_{\mathfrak{t}_2^i}-\zeta_{\mathfrak{t}_1^i}$, ... are independent and equally distributed with respect to the probability $\mathbf{P}_{i,0}$. From the representation of the process (ν_t,ζ_t) $(t\ge 0)$, given by section 46, it follows that $\mathbf{P}_{i,0}\zeta_{\mathfrak{t}_1^i}=\mu\mathbf{P}_{i,0}\mathfrak{t}_1^i$. Hence, $\mu\le 0$ implies $\mathbf{P}_{i,0}\zeta_{\mathfrak{t}_1^i}\le 0$ and, consequently, the greatest lower bound of the sequence $\zeta_{\mathfrak{t}_n^i}$ $(n=1,2,\dots)$ is almost certainly equal to $-\infty$ because it is a sequence of the sums of independent equally distributed variables with nonpositive means. Hence, $\mathbf{P}_{i,x}(\zeta_t\le 0$ for some $t\ge 0)=1$ for $x>0$, which can be written as $\mathbf{P}_{i,x}(\theta<\infty)=1$. By (87), this is equivalent to $Q\mathbf{1}=\mathbf{0}$.

If the conditions of the theorem are satisfied then $\mathbf{P}_{i,x}(\mathfrak{m}<\infty)=1$ and it follows from (169) and (171) with $s=0$ and $x=0$ that:

$$\|\mathbf{P}_{i,0}(\mathfrak{n}_{\mathfrak{m}}=j)\|_{i,j=1}^d=E-R(z)\{\mathfrak{B}R(z)\}^{-1}B(Q). \tag{172}$$

Because the matrix, A, is indecomposable, the matrix, Q, is thus also indecomposable, which according to (117) and (163), implies that both the matrix, $B(Q)$, and the matrix in the right hand side of (172) are indecomposable.

Because $\mathfrak{z}_{\mathfrak{m}}=0$, the Markovian chain imbedded into the subsequent stopping times, $\mathfrak{m}$, is ergodic and, if q_i is the stationary probability of the state $(i,0)$ where $\mathfrak{q}=(q_i)_{i=1}^d$, then:

$$\mathfrak{q}=\frac{1}{\langle \mathrm{e},\mathfrak{B}R(z)R(z)^{-1}\mathbf{1}\rangle}[\mathfrak{B}R(z)R(z)^{-1}]^{\mathsf{T}}\mathrm{e}. \tag{173}$$

According to Theorem 40, the left hand side of (165) is almost certainly (with respect to $\mathbf{P}_{i,x}$) equal to:

$$\mathbf{P}_{q,0}\int_0^{\mathfrak{m}} f(\mathfrak{n}_t,\mathfrak{z}_t)dt \Big/ \mathbf{P}_{q,0}\int_0^{\mathfrak{m}} g(\mathfrak{n}_t,\mathfrak{z}_t)dt,$$

where $\mathbf{P}_{q,0} = \sum_i q_i \mathbf{P}_{i,0}$.

To finish the proof, it is sufficient to verify that:

$$\mathbf{P}_{q,0}\int_0^{\mathfrak{m}} e^{q\zeta_t}v(\mathfrak{n}_t)dt = \frac{-1}{\langle \mathbf{e}, \mathfrak{B}R(z)R(z)^{-1}\mathbf{1}\rangle}\sum_{i=1}^{d} v(i)\left\{b_i e_i + \int_0^{\infty}\pi_i(x)e^{px}dx\right\} \quad (174)$$

for p ($p < 0$) having a sufficiently large absolute value.

It follows from the identity;

$$\int_0^{\infty} e^{px}\mathfrak{B}R(x)dx = B(p)A(p)^{-1},$$

which is implied by (85) and (164), that the right hand side of (174) is equal to:

$$\frac{-1}{\langle \mathbf{e}, \mathfrak{B}R(z)R(z)^{-1}\mathbf{1}\rangle}[\langle \mathbf{e}, B_2\mathbf{v}\rangle - \langle \mathbf{e}, B(p)A(p)^{-1}\mathbf{v}\rangle], \quad (175)$$

where $\mathbf{v} = (v(i))_{i=1}^{d}$.

To calculate the left hand side of (174), it is convenient to start by calculating the variables given by:

$$u(i,x) = \mathbf{P}_{i,x}\int_0^{\mathfrak{m}} e^{st}e^{p\zeta_t}v(\mathfrak{n}_t)dt, \qquad s<0, \qquad p\le 0.$$

According to the formula for complete probability;

$$u(i,x) = \mathbf{P}_{i,x}\int_0^{\tau_z} e^{st}e^{p\zeta_t}v(\mathfrak{n}_t)dt + \mathbf{P}_{i,x}[e^{s\tau_z}h(\mathfrak{n}_{\tau_z},\mathfrak{z}_{\tau_z})], \quad (176)$$

where;

$$h(j,y) = \mathbf{P}_{i,y}\int_0^{\tau} e^{st}e^{p\mathfrak{z}_t}v(\mathfrak{n}_i)dt.$$

The function, h, is a unique solution in $\mathbb{C}$ of the equation;

$$\mathfrak{A}h(i,x) + sh(i,x) = e^{px}v(i), \qquad x>0, \qquad h(i,0)=0. \quad (177)$$

A direct calculation demonstrates that:

$$\mathbf{h}(x) = (h(i,x))_{i=1}^{d} = [e^{xQ(s)} - e^{xp}E]\{A(p)+sE\}^{-1}\mathbf{v}. \quad (178)$$

Consequently, $h \in \mathbb{C}_2$ and Lemma 63 can be used, which together with (168), (176), and (177) demonstrates that:

$$\mathbf{u}(x) = (u(i,x))_{i=1}^{d} = \mathbf{h}(x) + R_s(z-x)\mathbf{c}$$

where the vector $\mathbf{c}$ does not depend on x.

By using Lemma 60 and the relation $\mathfrak{G}\mathbf{u}(0)+s\mathbf{u}(0) = -\mathbf{v}$, the following equation is obtained for the vector $\mathbf{c}$:

$$[B(Q(s)) - B(p)]\{A(p) + sE\}^{-1}\mathbf{v} + [\mathfrak{B}R_s(z) + sB_2R_s(z)]\mathbf{c} = -B_2\mathbf{v}.$$

This equation, and (178), yields:

$$\begin{aligned} \mathbf{u}(x) &= [e^{xQ(s)} - e^{xp}E]\{A(p) + sE\}^{-1}\mathbf{v} - R_s(z-x)\{\mathfrak{B}R_s(z) + sB_2R_s(z)\}^{-1} \\ &\quad \times\{[B(Q(s)) - B(p)][A(p) + sE]^{-1} + B_2\}\mathbf{v}. \end{aligned} \quad (179)$$

For $s \uparrow 0$ and $x = 0$, this becomes:

$$\left(\mathbf{P}_{i,0}\int_0^{\mathfrak{m}} e^{p\zeta_t}v(\mathfrak{n}_t)dt\right)_{i=1}^{d} = -R(z)\{\mathfrak{B}R(z)\}^{-1}\{[B(Q) - B(p)]A(p)^{-1} + B_2\}\mathbf{v}$$

for those $p < 0$ for which the matrix, $A(p)$, is invertible.

If the inner product of both sides of this equality with the vector of scalar probabilities given by (173) is taken, then expression (175) is obtained. This finishes the proof of the theorem.

65. If $\mu < 0$ and the measures in boundary condition (114) have a finite initial time, then the invariant measure of the process, $(\mathfrak{n}_t, \mathfrak{z}_t)$ $(t \geq 0)$, is finite. An expression for its complete mass can be considerably simplified if the notation, C^{-1}, is extended to the case where the matrix, C, is not invertible. The matrix, C^{-1}, is defined as follows. Let Π be the matrix of the projection onto the subspace of the vectors on which C is zero, i.e. $\Pi\mathbf{x} = \mathbf{x}$ if $C\mathbf{x} = \mathbf{0}$ and $\langle \mathbf{x} - \Pi\mathbf{x}, \mathbf{y}\rangle = 0$ if $C^{\mathsf{T}}\mathbf{y} = \mathbf{0}$. Then, the vector $\mathbf{y} = C^{-1}\mathbf{x}$ is a unique solution of the equation $c\mathbf{y} = \mathbf{x} - \Pi\mathbf{x}$ with the additional condition, $\Pi\mathbf{y} = \mathbf{0}$. Hence, $CC^{-1} = C^{-1}C = E - \Pi$.

Theorem. *If* $\mu < 0$ *and:*

$$\sum_{i=1}^{d}\left[\int_0^\infty x m_i(dx) + \sum_{j\neq i}\int_0^\infty x m_{ij}(dx)\right] < \infty,$$

then:

$$\sum_{i=1}^{d}\int_0^\infty \pi_i(x)dx = -\frac{1}{\mu}\langle \mathbf{e}, [B' - BA^{-1}A']\mathbf{1}\rangle,$$

where;

$$B' = B'(0) = \int_0^\infty xM_0(dx) + B_1 + \int_0^\infty xM_1(dx),$$

$$B = B(0) = B_0 + M_0[0,\infty), \qquad A = A(0) = A_0 + L_0[0,\infty),$$

$$A' = A'(0) = \int_0^\infty yL_0(dy) + A_1 + \int_1^\infty (y-1)L_1(dy).$$

Proof. For $x = 0$, $p = 0$, and $s < 0$, equation (179) becomes:

$$\begin{aligned}\left(\mathbf{P}_{i,0}\int_0^m e^{st}v(n_t)dt\right)_{i=1}^{d} &= -R_s(z)\{\mathfrak{B}R_s(z) + sB_2R_s(z)\}^{-1} \\ &\quad\times\{[B(Q(s)) - B](A + sE)^{-1} + B_2\}\mathbf{v}. \quad (180)\end{aligned}$$

The limit of this expression, as $s \uparrow 0$, is equal to:

$$-R(z)\{\mathfrak{B}R(z)\}^{-1}\left\{[B(Q) - B]A^{-1}\left(E - \frac{1}{\mu}A'\mathbf{1}\otimes\rho\right) - \frac{1}{\mu}B'\mathbf{1}\otimes\rho + B_2\right\}\mathbf{v} \quad (181)$$

Indeed;

$$\begin{aligned}[B(Q(s)) - B]Q(s)^{-1} &= B_1 + \int_0^\infty dy\int_y^\infty M(dx)e^{(x-y)Q(s)} \\ &\underset{s\uparrow 0}{\to} B_1 + \int_0^\infty dy\int_y^\infty M(dx)e^{(x-y)Q} = W, \quad (182)\end{aligned}$$

where $M(dx) = M_0(dx) + M_1(dx)$. Because the elements of the matrix, $M(dx)$, have finite initial moments then the elements of the matrix, W, are finite. Moreover, $[B(Q) - B] = WQ$. However, $QQ^{-1} = E - bf\mathbf{1}\otimes\mathbf{r}$, where $Q^\top\mathbf{r} = \mathbf{0}$ and $\langle\mathbf{r},\mathbf{1}\rangle = 1$. Hence, $W = [B(Q) - B]Q^{-1} + W\mathbf{1}\otimes\mathbf{r}$. It follows from (183) that $W\mathbf{1} = B'\mathbf{1}$ and hence;

$$W = [B(Q) - B]Q^{-1} + B'\mathbf{1}\otimes\mathbf{r}.$$

The matrix;

$$-QA^{-1} + \frac{1}{\mu}(QA^{-1}A' + E)\mathbf{1}\otimes\rho, \quad (183)$$

is the inverse of the matrix $K = -\lim_{s\uparrow 0}(A + sE)Q(s)$.

Indeed, the second identity of (94) gives:

$$K = A' + A_2Q + \int_0^\infty dy\int_y^\infty L(dx)[e^{(x-y)Q} - E]. \quad (184)$$

Moreover, $-KQ = A$. So, $K = -AQ^{-1} + K\mathbf{1}\otimes\mathbf{r}$ and, according to (184) and Lemma 63, $K\mathbf{1} = A'\mathbf{1}$, i.e.

$$K = -AQ^{-1} + A'\mathbf{1}\oplus\mathbf{r}.$$

If this matrix is multiplied by matrix (183) and the fact that $\langle\rho, A'\mathbf{1}\rangle = \mu$ is used, then matrix (183) is the inverse of K.

Therefore;

$$\lim_{s\uparrow 0}[B(Q(s)) - B](A + sE)^{-1} = -WK^{-1}$$

$$= [B(Q) - B]A^{-1}\left(\frac{1}{\mu}A'\mathbf{1} \otimes \rho\right) - \frac{1}{\mu}B'\mathbf{1} \otimes \rho$$

and, consequently, the limit of the right hand side of (180), as $s \uparrow 0$, is vector (181), which, if $\mathbf{v} = \mathbf{1}$, is equal to:

$$-R(z)\{\mathfrak{B}R(z)\}^{-1}\left\{-\frac{1}{\mu}[B(Q) - B]A^{-1}A' - \frac{1}{\mu}B' + B_2\right\}\mathbf{1}.$$

By using (173), the following is obtained:

$$\mathbf{P}_{q,0}\mathfrak{m} = \frac{-1}{\langle \mathbf{e}, \mathfrak{B}R(z)R(z)^{-1}\mathbf{1}\rangle}\left\{\langle \mathbf{e}, B_2\mathbf{1}\rangle - \frac{1}{\mu}\langle \mathbf{e}, [B' - BA^{-1}A']\mathbf{1}\rangle\right\},$$

which, together with (174) proves the theorem.

66. If $d = 1$, then the process, ζ_t $(t \geq 0)$, is a process with independent increments and, formula (104), with $f = 1$ and $x = 0$, gives the following:

$$\begin{aligned} \mathbf{P}_0 e^{s\theta_z} &= 1 - sR_s(z)\frac{1}{Q(s)} - \int_0^z R_s(y)dy \\ &= 1 - \frac{s}{Q(s)}R_s(z) - s\int_0^z R_s(y)dy, \end{aligned} \tag{185}$$

where $\int_0^\infty e^{px}R_s(x)dx = 1/(A(p)+s)$, $p < Q(s)$, $A(Q(s)) + s = 0$, $\mathbf{P}e^{p(\zeta_t - \zeta_0)} = e^{tA(p)}$ and $A(p) = a_1 p + a_2 p^2 + \int_0^\infty [e^{py} - 1 - p(y \wedge 1)]l(dy)$.

If, moreover, $\mu = A'(0) < 0$, then $Q(0-) = 0$ and, consequently;

$$\frac{s}{Q(s)} = -\frac{A(Q(s))}{Q(s)} \underset{s\uparrow 0}{\longrightarrow} -\mu.$$

By taking the limit of (185), then;

$$-\mu R(z) = \mathbf{P}_0(\theta_z = \infty) = \mathbf{P}(\sup_{t\geq 0} \zeta_t - \zeta_0 < z).$$

Boundary condition (114), for $d = 1$, becomes:

$$b_2\mathfrak{A}f(0) = b_1 f'(0) + \int_0^\infty m(dy)[f(y) - f(0)], \tag{186}$$

and condition (163) implies that $b_1 + m(0,\infty) > 0$.

Theorems 62 and 65, in this particular case, give an expression for the stationary distribution of the Markovian process, $\mathfrak{z}_t$ $(t \geq 0)$, in terms of the distribution of the absolute maximum of the process with independent increments, ζ_t $(t \geq 0)$.

Corollary. *If $d = 1$, $\mu < 0$, and $\int_0^\infty xm(dx) < \infty$, then:*

$$\lim_{t\to\infty} \frac{1}{t}\int_0^t g(\mathfrak{z}_u)du = \int_0^\infty g(x)dF(x),$$

where;

$$\int_0^{\infty} |g(x)|dF(x) < \infty,$$

$$F(x) = \frac{-\mu b_2 + b_1 G(x) + \int_0^x G(x-y)m(y,\infty)dy}{-\mu b_2 + b_1 + \int_0^{\infty} y m(dy)},$$

$$G(y) = \mathbf{P}(\sup_{t\geq 0} \zeta_t - \zeta_0 < y).$$

In a particular case, where $b_2 = m(0,\infty) = 0$ in boundary condition (186), the stationary distribution of the process, $\mathfrak{z}_t$ $(t \geq 0)$, is equal to the absolute maximum of the process, ζ_t $(t \geq 0)$. This result is similar to the well known fact from queueing theory that the stationary distribution of the queue length in a single server waiting system with an exponential service time, which has a nonstandard Poisson flow of inquiries, coincides with the distribution of the absolute maximum of the process with independent increments, which describes the behavior of the system with a nonempty queue.

67. Let, as before, $d = 1$. The process, $\mathfrak{z}_t$ $(t \geq 0)$, is continuous if and only if $A(s) = a_2 s^2$ and the measure, m, from boundary condition (186) is equal to zero. In other words, the process with independent increments, ζ_t $(t \geq 0)$, is a Wiener process (up to a constant positive factor) and the boundary condition, by using (163), can be written as $b_2 f''(0) = f'(0)$. In this case $\int_0^{\infty} e^{px} R(x)dx = 1/(a_2 p^2)$ and, consequently, $R = x/a_2$. By Theorem 62, the atom of the invariant measure at zero is equal to b_2 and its density away from the origin is constant and equal to $1/a_2$. Hence, if the process, ζ_t $(t \geq 0)$, is continuous, then:

$$\lim_{t\to\infty} \frac{\int_0^t f(\mathfrak{z}_u)du}{\int_0^t g(\mathfrak{z}_u)du} = \frac{b_2 f(0) + \frac{1}{a_2}\int_0^{\infty} f(x)dx}{b_2 g(0) + \frac{1}{a_2}\int_0^{\infty} g(x)dx} \tag{187}$$

is almost certainly true if the functions f and g are absolutely integrable and the denominator is not zero.

If the boundary condition is of the form (186), then the atom at zero is equal to b_2 and the density of the invariant measure away from the origin is:

$$\frac{1}{a-2}\left\{b_1 + x m(x,\infty) + \int_0^x y m(dy)\right\} = \frac{b_1}{a_2} + \frac{1}{a_2}\int_0^x m(y,\infty)dy.$$

In the case when the process, ζ_t $(t \geq 0)$, is continuous and the boundary condition is of the form (186), the limit in the left hand side of (187) is equal to:

$$\frac{a_2 b_2 f(0) + b_1 \int_0^{\infty} f(x)dx + \int_0^{\infty} f(x)\int_0^x m(y,\infty)dydx}{a_2 b_2 g(0) + b_1 \int_0^{\infty} g(x)\int_0^x m(y,\infty)dydx}$$

under the usual existence conditions for the integrals.

68. An explicit calculation of the invariant measure is possible for $d = 1$ when ζ_t $(t \geq 0)$ is a stable process with independent increments such that $A(s) = |s|^\alpha$, $1 < \alpha < 2$, and the boundary condition is $b_2 \mathfrak{U} f(0) = \int_0^\infty [f(y) - f(0)] y^{-\beta} dy$, where $1 < \beta < 2$. The formula;

$$|s|^\alpha = \frac{\alpha(\alpha-1)}{\Gamma(\alpha)} \int_0^\infty (e^{sx} - 1 - sx) x^{-\alpha-1} dx, \qquad s \leq 0,$$

where $\Gamma(\alpha) = \int_0^\infty x^{-\alpha} e^{-x} dx$ is Euler's gamma function, demonstrates that the process, ζ_t $(t \geq 0)$, satisfies condition (84) and, because the integral $\int_0^\infty y^{-\beta} dy$ diverges, condition (117) is valid. Further, $\int_0^\infty e^{px} R(x) dx = |p|^{-\alpha}$. Hence, $R(x) = x^{\alpha-1}/\Gamma(\alpha)$.

In this case the parameter, μ, is equal to zero, the atom of the invariant measure at zero is equal to b_2 and its density away from the origin is given by:

$$\frac{1}{\Gamma(\alpha)} \int_0^\infty [x^{\alpha-1} - (x-y)^{\alpha-1}[y^{-\beta} dy + \frac{x^{\alpha-1}}{\Gamma(\alpha)} \int_x^\infty y^{-\beta} dy$$
$$= \frac{x^{\alpha-\beta}}{\Gamma(\alpha)} \left[\frac{1}{\beta-1} + \int_0^2 \frac{1-(1-y)^{\alpha-1}}{y^\beta} dy \right].$$

In particular, if $\alpha = \beta$, the density of the invariant measure away from the origin, similar to Section 67, is constant.

Hence, if $\int_0^\infty \{|f(x)| + |g(x)|\} x^{\alpha-\beta} dx < \infty$ and $\int_0^\infty g(x) x^{\alpha-\beta} dx \neq 0$, then:

$$\lim_{t\to\infty} \frac{\int_0^t f(\mathfrak{z}_u) du}{\int_0^t g(\mathfrak{z}_u) du} = \frac{c b_2 f(0) + \int_0^\infty f(x) x^{\alpha-\beta} dx}{c b_2 g(0) + \int_0^\infty g(x) x^{\alpha-\beta} dx}$$

is almost certainly true, where:

$$\frac{1}{c} = \frac{1}{\Gamma(\alpha)} \left[\frac{1}{\beta-1} + \int_0^1 \frac{1-(1-y)^{\alpha-1}}{y^\beta} dy \right].$$

It should be remarked that, if $\alpha < \beta$, then the value of the invariant measure of each of the intervals $(0,1)$ and $(1,\infty)$ is infinite.

69. Condition (84) eliminates the possibility of $\zeta_t = \zeta_0 - t$ (when, as before, $d = 1$). However, Theorem 62 still remains valid. The proof of this is left as an exercise. The boundary condition for the process, $\mathfrak{z}_t$ $(t \geq 0)$, must be:

$$-b_2 f'(0) = \int_0^\infty [f(y) - f(0)] m(dy)$$

with, by (163), a nonzero measure, $m(dy)$. Because $A(s) = -s$ then $R(x) = 1$ and, consequently, the density of the invariant measure away from the origin is equal to $m[x,\infty)$, and its atom at zero is b_2. Hence, if $\int_0^\infty x m(dx) < \infty$, and $\int_0^\infty |f(x)| m(x,\infty) dx < \infty$, then:

$$\lim_{t\to\infty} \frac{1}{t} \int_0^t f(\mathfrak{z}_u) du = \frac{b_2 f(0) + \int_0^\infty f(x) m(x,\infty) dx}{b_2 + \int_0^\infty x m(dx)}$$

is almost certainly true, which agrees with Corollary 66. Indeed, in this case, $\sup_{t\geq 0}(\zeta_t - \zeta_0) = 0$ with unit probability. Hence, with the notation used in Corollary 66, $G(y) = 1$ for $y > 0$, and thus;

$$F(x) = \frac{b_2 + \int_0^x m(y,\infty)dy}{b_2 + \int_0^\infty ym(dy)}.$$

70. Another case, which does not satisfy condition (84) and which has a simple explicit expression for the invariant measure, is the case of a Poisson process, ζ_t $(t \geq 0)$, with negative drift, zero mean value and exponentially distributed positive jumps i.e.

$$A(s) = a_1 s + c\int_0^\infty (e^{sx} - 1)\lambda e^{-\lambda x}dx,$$

where $\lambda > 0$, $G > 0$, and $\lambda a_1 + G = 0$.

Theorem 62 remains valid in this case also (the proof is left as an exercise).

With such assumptions, $A(s) = \dfrac{cs^2}{\lambda(\lambda - s)}$ and, consequently, $R(x) = \dfrac{\lambda}{c} + \dfrac{\lambda^2 x}{c}$. The boundary conditions can always be written with $b_1 = 0$, i.e.

$$b_2\mathfrak{A}f(0) = \int_0^\infty [f(y) - f(0)]m(dy).$$

The density of the invariant measure away from the origin is given by:

$$\frac{\lambda}{c}m(x,\infty) + \frac{\lambda^2}{c}\int_0^x m(y,\infty)dy$$

and, as usual, its atom at zero equals b_2.

Short Bibliographical Indications

Chapter 1

Section 1.1. Fundamental results of G.D. Birkhoff, A.Ja. Khintchine, E. Hopf are given. The proof of Lemma 5 belongs to A.M. Garsia.

Birkhoff, G.D. Proof of the ergodic theorem. *Proc. at. Acoc. Sci.*, 1931, **17**, p. 656–660.

Khintshine, A.Ja. Zu Birkhoff Lösung des Ergodenproblems. *Math. Ann.*, 1932, **107**, p. 485–488.

Hopf, E. Ergodentheorie. *Ergeb. Math.*, 1937, **2**.

Section 1.2. Stationary processes were introduced by A.Ja. Khintchine. Theorem 12 is close to the ergodic theorem from the paper of P. Franken and A. Shtreller. Material of this paper is used also in Section 14. Results of other sections are not highlighted in journals.

Khintshine, A.Ja. Korrelationstheorie der stationären stochastischen prozesse. *Math. Ann.*, 1934, **109**, p. 604-615.

Franken, P. and Shtreller, A. Stationary generalized random processes. *Theory of Probab. and Appl.*, 1979, **24**, p. 78–90.

Section 1.3. Notion of permutative random variables belongs to B. de Finetti. Connection of ergodic properties with the existance of moments of regeneration was found by W. Feller. W.L. Smith gave contemporary definition of regenerating process. B.A. Sevastjanov indicated ergodic properties of branching processes with immigration

B. de Finetti. Funzione caratteristica di un fenomeno aleatorio. *Memo. R. Accoc. Naz. Lincei, 6 série*, 1930, **4**, p. 86.

Smith, W.L. Renewal theory and its ramifications. *J. Roy. Stat. Soc.*, 1958, **20**, p. 243–284.

Feller, W. Fluctuation theory of recurrent events. *Trans. Amer. Math. Soc.*, 1949, **67**, p. 98–119.

Sevastjanov, B.A. Limit theorems for branching random processes of special kind.

Theory of Probab. and Appl., 1957, **2**, p. 339–349. (In Russian).

Chapter 2

Section 2.1. The proof of Theorem 33 is similar to the work of T.E. Harris.
Harris, T.E. The existence of stationary measures for certain Markov processes. In: *Proc. Third Berkeley Symp. Math. Stat. and Probab.*, 1956, vol. **2**, p. 113–124.

Section 2.2. Section contains revised and supplemented paper of the author and I.I. Ezhov.
Ezhov, I.I. and Shurenkov, V.M. Ergodic theorems connected with Markovian property of random processes. *Theory of Probab. and Appl.*, 1976, **21**, p. 635–638.

Chapter 3

Section 3.1. Subsections 45, 46 and 47 were written by means of the material of the first part of I.I. Ezhov and A.V. Skorokhod's paper. Equation (92) is given in the second part of this paper. Other results of the section are partly highlighted in joint paper of V.S. Korolyuk and the author.
Ezhov, I.I. and Skorokhod, A.V. Markov processes, which are homogeneous with respect to second component. I. II. *Theory of Probab. and Appl.*, 1969, **14**, p. 3–14, 679–692.
Korolyuk B.S. and Shurenkov V.M. Potential method in boundary problems for random walks on Markov chain. *Ukrain. Math. Journal*, 1977, **29**, p. 464–471.

Sections 3.2 and 3.3. Material of these two sections is based on the paper of the author.
Shurenkov V.M. Bounary problems and ergodic theorem for processes with independent increments. *Theory of Probab. and Appl.*, 1979, **24**, p. 52–61.

Bei Fragen zur Produktsicherheit wenden Sie sich bitte an:
If you have any questions regarding product safety,
please contact:

Walter de Gruyter GmbH
Genthiner Straße 13
10785 Berlin
productsafety@degruyterbrill.com